P9-AFV-128

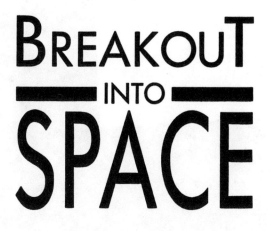

BREAKOUT
═ INTO ═
SPACE

Mission
for a
Generation

GEORGE HENRY
ELIAS

William Morrow and Company, Inc.
New York

Recognizing the importance of preserving what has been written, it is the policy of William Morrow and Company, Inc., and its imprints and affiliates to have the books it publishes printed on acid-free paper, and we exert our best efforts to that end.

Library of Congress Cataloging-in-Publication Data

Elias, George Henry.
    Breakout into space : mission for a generation / George Henry
Elias.
        p. cm.
    ISBN 0-688-07703-X
    1. Outer space—Exploration. I. Title.
TL790.E44 1990
919.9'04—dc20   89-12169
                                                            CIP

Printed in the United States of America

First Edition

1 2 3 4 5 6 7 8 9 10

BOOK DESIGN BY WILLIAM McCARTHY

This book is dedicated
to those
men, women, and children
who one day
will leave this solar system
on the first starship

MAY THEY FIND THE NEW EARTH

Almost thirty years have passed since fifty-nine young political activists wrote *The Port Huron Statement*. This manifesto—recently called "one of the pivotal documents in postwar American history"—had a profound influence over the politics of the 1960's, and helped make a new organization called Students for a Democratic Society (SDS) nationally prominent. A subtitle of *The Port Huron Statement* is "Agenda for a Generation." Its authors were presenting their ideas to their contemporaries as "an effort in understanding and changing the condition of humanity in the late twentieth century."

We need another agenda for this generation, for the 76 million Americans born between 1946 and 1964. Those of us born during those years now form about 60 percent of the electorate. We are now responsible for our country, and we have the power to decide its direction for decades. But we have not yet chosen our course, and the nation totters forward guided by political ideas formed a half-century ago. Meanwhile, global problems are multiplying. World population races ahead, nuclear weaponry accumulates, and ecological disorders become more frequent. We need a new public philosophy.

Some say that our generation does not exist as a political force. They believe that we are too diverse to unite behind a common program. They argue that the decisive influence we exercised over national affairs in the 1960's was circumstantial and unlikely to be repeated. I oppose this view. I believe that—despite the importance of our contributions to the civil-rights and antiwar movements—we will benefit the nation far more in our political maturity than we did in the 1960's. The 1990's will be the beginning of a political activism that will eclipse the juvenile efforts of our youth. Of course, we have our differences. We are divided by age, gender, race, education, income, geography, and marital status. But such is always the case before leaders do their work of transforming a divided constituency into a united people.

My primary purpose in writing this book is to put the issues involved in the settlement of space into the mainstream of American politics. Opening the frontier above us for millions of people is a moderate, sensible course, given the heritage of this country and the problems of this planet. Making space accessible to ordinary people is the key to our survival both as a nation and as a species. The United States is a country formed by over three hundred years of frontier experience. We have neither the national character nor institutions to prosper on a closed planet of fiercely competitive rivals. Let us do what we do well—explore, settle, and develop—and leave to other nations the rewards of foreign entanglements.

The construction of an interplanetary civilization is essential to the survival of our species. The history of life is the history of growth and expansion. After one ecological niche is filled, life moves on to fill another. We humans now occupy an entire planet, and it is time for us to move on to a larger habitat. An empty solar system awaits us. Earth's rising population and expanding ecological dilemmas are signs that we have overstayed our welcome here. We are making our home planet sick. Let us have the courage to do what life has always done. Let us obey natural law and move on.

The task of our gifted generation is to make space accessible to millions and to begin building a civilization that spans the solar system. Our tens of millions of votes make us America's rulers. And no other ruling class today has the education, longevity, political skills, and national resources that we possess. Should we fail in this endeavor, we face an unhappy and perhaps impoverished old age, passing on to our children a patrimony much smaller than what we received. But if we do what is necessary, we shall have found a challenge worthy of our gifts, we shall have performed a great service to humanity, and we shall be rewarded far more richly than we could ever imagine now.

I am happy to acknowledge the assistance I received in completing this work.

First, I wish to acknowledge a heavy debt to several people whose literary skills and candor in reviewing my work were indispensable: Paul Brumbaum, David Hans, Juan Hovey, and the Reverend John Turpin. I thank you again for the time and assistance you gave me.

Second, I wish to thank my agent, Michael Larsen, and my editors, Jim Landis and Margaret Talcott of William Morrow & Company, Inc. Without Mr. Larsen's lucid handbook, *How to Write a Book Proposal*, and his early recognition of promise in the seeds of my work, this book might never have been published. I also wish to commend Ms. Talcott for her excellent editorial work and Mr. Landis for his patient understanding after I had missed numerous deadlines.

Next, I wish to thank those who read my manuscript at various times. The perspective they gave me was invaluable: Joseph Libs, Rod McLeod, David Riggs, Harrison Sheppard, John (J. T.) Taylor, the Reverend Jim Kitchens, the Reverend Rebecca Kuiken, Scott Bottles, John Carr, Laird Durley, John and Lisa Elias, Denise Fort, Grant Franks, Faye Beverett, Rob and Nancy Jolda, Gregg Maryniak, Marcia Mathog, David Vogel, and Pat Walker. I also wish to thank Harry Yue, owner of Business Services Network, as well as John Wilson and Angelo Izzo, owners of Acquisition Management & Brokerage, for the use of their word-processing equipment. Finally, I thank the library staff at the Mechanics Institute in San Francisco for their assistance. Richard Bittner and Carla McLean were especially helpful.

Finally, I owe thanks to Karen Evjen—my onetime wife, lifetime friend, and mother of our dear daughter, Ingrid. Her unfailing humor and affection brightened many days in my long labors.

CONTENTS

BREAKOUT
INTO
SPACE

CHAPTER ONE

The Gifted Generation

The parents of those of us born between 1946 and 1964 form one of the most remarkable generations in American history. They began their lives in the bleak, black-and-white years of the Depression, when tens of millions suffered the deprivations and humiliations of poverty. But they survived and went on to fight World War II, a war that was larger than any ever fought. By the time victory was won, they had carried the nation to a pinnacle of international power and wealth that may never be regained. Their achievements in peace were no less memorable. They rebuilt Europe through the Marshall Plan, a plan that historians regard as one of the greatest triumphs of American diplomacy. They forged an economy that produced the greatest surge of growth and the most sustained prosperity that the world has ever seen. In 1969, they landed men on the Moon, perhaps opening a new era for all humanity. And finally, they created leaders—such as Martin Luther King, Jr., and the Kennedy brothers—whom we will not soon forget.

Our mothers and fathers may also be proud of their private lives. Their marriages are among the most stable and successful in our history. Grateful to have had families after the hazards of depression and war, they married for life and devoted themselves to raising children. Their efforts were rewarded. The generation they reared has become renowned as the largest, best-educated,

and most promising generation since North America was first settled nearly four hundred years ago. We who are members of this
group represent our parents' most valuable legacy to the future.
What can we say about our generation?

GIFTS

We are a gifted generation. During our childhood and adolescence,
we received benefits that neither our ancestors nor most of our
contemporaries enjoyed. War, poverty, oppression, and backward-
ness blight most lives. Few have the peace, prosperity, liberty, or
technology that formed just part of our own inheritance. We had
these precious gifts and more. We also had families that seemed to
exist for our sake and an opportunity for an education that our
forebears never had. Of course, we grew up unconscious of how
singularly we were privileged. The fortunate often assume that
their blessings are natural and ordinary.

In many ways, the America of our youth was a golden age.
The nation was at peace. Great prosperity prevailed. New technology gave us leisure and broadened the scope of our lives. Neither the tyranny of a dictator nor the restrictions of an oppressive
society affected our behavior. We were free to do what we wanted,
and there was more to do than ever before.

A great war had just ended, and the world was weary of
fighting. No one bombed our cities or shot our fathers. We did not
live in fear of sudden death or slow torture at the hands of enemies.
Memories of senseless violence did not scar our minds. Unlike
Lebanese or Irish children today, we did not spend our time in the
schoolyard playing war in deadly imitation of adults. On the contrary, we grew up in a time of peace, despising the military. In the
1950's, America had fought no wars within its borders for nearly
100 years, and the last foreign invasion had occurred almost 150

years before. War seemed as far from our sunny backyards and quiet suburban streets as the barbarian invasions of the Roman Empire.

Our country was a land of plenty, in contrast to almost every nation in history. Food in colorful packages crowded supermarket shelves, competing for our attention. If we needed clothes, our parents bought them; and if fashions changed at school, they bought us more. We lived in new homes that had picture windows, front lawns, and backyards with gardens, playgrounds, and barbecues. We always had the toys we wanted (bicycles, dolls, games, playhouses, or skates). And at the end of the year, our parents piled presents for us high around the Christmas tree. Living in an affluent society, we had many superfluities and lacked no necessities. While other people scrabbled for bare existence, we had more than we needed.

Modern technology may have been the most distinctive element in postwar America. Advances in communication, transportation, and home appliances separated us both from previous generations and from our contemporaries in less modern countries. Television increased our experiences and our information a hundredfold. No longer would we have a provincial outlook and be ignorant of what was happening in the world. Cars and air travel had similar effects. They freed us from our neighborhoods and our regions, enlarging greatly the space and scope of life. Home appliances also freed us, by giving us and our parents leisure for other activities. Technology had automated the home, and we were no longer drudges bound to a household economy struggling for basic amenities. Because of these changes, the world in which we spent our youth was not the world of our parents and our grandparents. The pace, scale, and extent of our lives was greater than theirs. Life's possibilities had widened.

The liberties society allowed us set us more apart from other generations. This freedom derived from two sources. First, we were citizens of a nation with a heritage of civil liberties. As Americans, we inherited the rights prescribed in the Declaration of Independence, the Constitution, and the Bill of Rights. Few governments ever grant such freedom to the governed. Second, we lived in an era in which traditional authorities, such as church and family, had lost much of their power. To a surprising extent, the young regulated their own behavior. The social history of those years was the story of the breaking and discarding of rule after rule. We decided how we would interact with our peers and with the opposite sex; we decided what we would read and what we would watch; we decided what we would study and what careers we would pursue. Rarely have the young been left more completely to themselves.

Peace, abundance, technology, and freedom were gifts that graced our childhood. We did not have to fight for our lives or for our bread. Nor were we ignorant of the world, or provincial in experience. Neither daily chores nor an oppressive society enslaved us. It is no wonder that our parents felt proud of us. They bestowed on us a kind of life that few have ever enjoyed, one that in its privileges rivaled aristocracies of the past.

We had two additional advantages: family and education. In contrast to the generations preceding us—whose parents were distracted by the struggle for existence—we grew up in families devoted to raising us. Because of the deprivation they had suffered in their childhoods, and because of the delay in their lives occasioned by the war, our parents were determined to give us everything they had missed. We represented the culmination of their hopes. And their greatest hope centered on giving us an education. When our parents were young, education was a privilege of the upper classes. Many believed that it offered a kind of secular salvation, revealing a path to everything good that life contained. Had they not been able to give us an education, our parents would have regarded themselves as failures.

Obsessed with raising children who enjoyed all the social

advantages that they had missed, our parents devoted themselves to us. The Depression and the Second World War had deformed their own lives by consuming their youth, bottling up their dreams for fifteen years and uprooting them from their homes. When their opportunity to start families finally came in the mid-1940's, they married for life and raised the most privileged generation this country has ever seen. Reinforced by books like Dr. Benjamin Spock's *Baby and Child Care*, they made children and family a religion. Life became parenthood and revolved around the family. Nothing was too good for us.

Not only were we part of families that stayed together, but we were also the first and the last generation to have the benefit of full-time mothers. Overwhelmed by the manual labor of their unautomated households, women in earlier generations had little time to invest in the development of their children. The newer generations face a similar circumstance. Mothers holding full-time jobs have little leisure to spare. But the childhood of those of us born in the postwar generation was different, because we had the luxury of a loving adult who stayed home to watch over us as we negotiated the hazards of youth. Educated and intelligent women spent their lives caring for their children. We had a parent at home during the crucial preschool years, and very few of us were ever latchkey children returning to empty houses after school. Whether the prevalence of full-time motherhood was good for either women or society is not the issue here. What is significant is that we, the children of these mothers, benefited from the continual presence of a caring adult in our homes. And it was an advantage that we share with neither our parents nor our children nor, possibly, our grandchildren.

Americans have always valued education highly, and no generation born in this country has received a better education than we have, we postwar children. In our youth, parents mobilized to build us classrooms, buy us the best textbooks, and hire the best teachers. Controversy shook the nation at any sign that our schooling was deficient, as when the launch of Sputnik in 1957 suggested that the USSR had better training in the sciences. By any statistical measure, we are better educated than any previous generation. Nearly 85 percent of those born between 1947 and 1951 finished high school, although only 38 percent of their parents did so. The fathers of two thirds of today's college graduates never went beyond high school. We are twice as likely as our parents to have attended college, and three times as likely to have done so as our grandparents. Almost 44 percent of those born between 1946 and 1960 have had at least four years of college. No generation has had as many college graduates as ours. The circumstances that permitted such a flowering and a harvest may never recur.

We are the gifted generation. Born at a peak of national power and prestige, we were the unconscious beneficiaries of centuries of philosophic genius, political heroism, economic ingenuity, and technical craft. We lived in the midst of a horizonless peace, while at our call stood the most productive economy known to man. Free from social and political oppression, nurtured in complete families, attended by full-time mothers, and educated to an unprecedented degree, we drank deeply from the cup of optimism that has always been an American birthright. Although our families were middle class, our privileged upbringing gave us a pride and self-confidence that was aristocratic, perhaps royal. Indeed, from an international and historical perspective, we may be as unique in our world as princes and princesses of past kingdoms were in theirs. It is as if we are the heirs of an ancient family of great dignity and wealth. We have numerous prerogatives and face spacious vistas. We have received much, very much, and from us great things are expected.

FAILURES

In 1967, *Time* magazine named our generation its "Man of the Year." We received an honor bestowed on only a handful of world leaders since the news magazine began making its annual selections in the 1930's. "Never have the young been so assertive or so articulate, so well educated or so worldly. This is not just a new generation, but a new kind of generation." In our lives—predicted the magazine—we would cure cancer and the common cold, create blight-proof and smog-free cities, help end racial prejudice, enrich the underdeveloped world, and end poverty and war.

On such words were we raised. "This is the generation," said Patrick Caddell, the famous pollster and political strategist, "that grew up believing it was going to reshape the world." From our earliest years, we have believed that we were somehow special and had a mission in life. "Of the ideas that bind" this generation, wrote Landon Jones, author of the widely quoted book *Great Expectations*, "none is stronger than the belief that [it] has a mission in life." We wanted not only the affluence in which we had lived as children, but also something beyond mere wealth. We wanted a perfect society and the perfect person. We wanted it all: the best education, national reform, dual-career marriages, good health, and fulfilled lives. But since we left school, our accomplishments in the economy, our families, and politics have fallen far short of these ideals. We have failed in many ways.

Failure in Work

Our first and most fundamental failure is economic. We are failing to match the standard of living of our parents. We are the first generation since this continent was colonized that has not advanced the nation's economic condition. The conventional American dream—one or two cars, vacation travel, family home, comfortable retirement, educated children—is becoming elusive to many of us.

When my family made the big leap to the suburbs twenty-two years ago, my father worried that his children did not fully appreciate this achievement. I remem-

ber vividly that he sat me down in our new home and
said pointedly: "You know, this is the nicest house I've
ever lived in." I'm sure it never occurred to him, and it
certainly never occurred to me until much later, that it
might be the nicest house I'd ever live in as well.

This generation "may never achieve the relative economic success
of the generations immediately preceding it or following it," declared the 1980 United States Budget.

Before 1973, the economy functioned admirably. Until that
year, living standards had risen 2.5–3.0 percent annually since the
end of World War II. From 1947 to 1973, family income had
nearly doubled; in fact, it had never gone three years without
setting a new record. During that period, men's real wages had
climbed about 110 percent between their twenty-fifth and thirty-
fifth years, and at least 30 percent between their fortieth and fiftieth
years. These times were prosperous.

But a quiet depression began in 1973—about the same time
that our generation started entering the economy in force. After
that last good year, income growth slowed dramatically, until by
the mid-1980's family income was lower than the 1973 high.
Instead of rising during their fortieth and fiftieth years, men's real
wages fell 14 percent; and they rose just a meager 16 percent for
men between their fast-track twenty-fifth and thirty-fifth years. A
leading Democratic senator states that since Europeans settled in
America nearly four hundred years ago, there may never have been
a similar period in which family income did not rise. "In the
decades prior to the 1970's," stated the economists Frank S. Levy
and Richard C. Michel in 1986, "children expected early on to
live better than their parents. Such is not now the case."

What had happened? Economists believe that the decline in
family and individual incomes was not caused by a major historical
event, such as the end of the Vietnam War, but by something much

more mundane—a drop in the growth of worker productivity. The source of rising living standards is an increasing output per man hour; because we are wealthier as a nation if there are more goods and services to distribute. But after 1973, worker productivity stopped growing. Each year, the private sector of the economy produced only marginally more per worker than the year before. Hence, family incomes stopped growing too.

Why did worker productivity stop growing? Although there are several theories, economists are not certain. According to one school of thought, growth in productivity ceased because of a rapid increase in energy prices, ten years of inflation, and an expanding labor force. These factors may have been influential; but by 1984, they were no longer important. Energy prices were stable, inflation low, and labor growing more slowly. Yet worker productivity remained disappointing. Obviously, other causes were at work.

The evidence suggests that we of this generation bear some of the responsibility for the decline in productivity growth and, consequently, for the decline in income growth. It appears to be too much of a coincidence that productivity began to deteriorate at the same time that we were joining the economy. Just as the 1960's were our college years, the 1970's were the years we began our careers. Our work habits must have had some effect, in the same way that the habits of our parents must have contributed to the prosperity of the postwar years. Their tendencies toward job commitment and hard work formed the foundation of the boom years. Our habits, on the other hand, may not have had such a wholesome effect. Long before corporate raiders began restructuring corporations and destroying job security, we were notorious for lacking loyalty to an employer. We job-hopped. We were also less responsive to authority and tended to regard our leisure activities as more important than our work lives. All these qualities may be meritorious, but workers do not increase their output if they are always learning a new job and if they pay scant heed to their employer's interests.

Failure in Private Life

Our most profound failure has been in family life. "This generation," wrote Landon Jones in *Great Expectations*, "has done more to undermine kinship than any previous generation." We have married less, married later, and had fewer children. Divorce among us has become epidemic, and we have abandoned millions of children. We are the first generation of parents to be widely unavailable to their young. As the fifties cult of the child has given way to the eighties cult of the adult, the idea of keeping the marriage together for the sake of the children has been discredited and the sense of responsibility of parent to child has eroded. We have broken links between husbands and wives, parents and children, grandparents and grandchildren. We seem to be doing all we can to lessen generational continuity and break the family's hold on the individual.

Divorce rates have skyrocketed. When we started forming families in the seventies, the divorce rate quickly doubled. By 1980, divorces for couples under thirty were nearly triple what they were in 1970. According to one estimate, 40 percent of the marriages initiated in 1980 would end in divorce—and the chances of success dropped at remarriage. Divorce rates continued to climb in the early eighties; by 1984, they were quadruple the 1960 rate for couples under thirty. Only at the end of the decade did the divorce rate fall, as we began to approach and pass our fortieth birthdays. But the damage had been done.

Children bear the brunt of divorce. Millions of our children will grow up in split families, and nearly half are likely to spend time living with only one parent before age eighteen. By the late 1980's, about one in four children lived in a single-parent home—

triple the percentage of 1950. What had been a deviant pattern had become ordinary.

We all know that children suffer acutely from divorce. They are often psychologically wounded, with feelings of loneliness, depression, and guilt. They experience loss, a heightened sense of impermanence, and a drop in their feelings of security. Their own marriages will tend to fail because they grow up without observing a marriage firsthand and with little idea of how a successful family operates. They may also suffer a much greater injury. The homicide rate for children under four tripled between 1950 and 1975; and by 1980, homicide was the greatest cause of childhood mortality.

A primary responsibility of any generation is the care and rearing of its children and the transmission of its culture to them. Certainly we who benefited so much from this fundamental human tradition have the most reason to perpetuate it. Yet we have not. We have left our spouses, abandoned our families, and neglected our children. For our parents—and perhaps most of the generations before us—family came first, marriage second, and self last. We reversed that order. We have moved the nation far from that condition that impelled Alexis de Tocqueville, the great observer of American life, to write, "Of all countries in the world America is the one in which the marriage tie is most respected."

Failure in Politics

Our most conspicuous failures have been political. We have failed to exercise an influence commensurate with what we have to offer. "[T]he apparent failure of the sixties generation to provide fundamental new direction to American politics," wrote Michael Delli Carpini, a political scientist at Rutgers University, in 1986, "is a cause for great disappointment." Although we have had some successes, they are either part of our youth—the civil-rights and antiwar movements—or part of larger trends that were perhaps inevitable anyway—the women's, consumer, and environmental movements. And in the few areas—such as social attitudes towards sex and drugs—where we have changed American society decisively, the results have not been benign. Because of our influence, recreational drugs became an acceptable part of middle-class life, and the social barriers to extramarital sex disappeared. Twenty years later, we see the consequences. Drug use is rampant, from high school to high society, and medical experts fear the consequences of AIDS, a sexual epidemic for which there is no known cure.

Let us not forget that we are the generation who spoke so loudly about the coming revolution. We were the ones who accused authorities of hypocrisy and demanded improvement. We were the ones who wanted to change the system. "We must name that system. We must name it, describe it, analyze it, understand it and change it," proclaimed Paul Potter, president of the Students for a Democratic Society (SDS), in early 1965. At its peak, the SDS led a mass movement with hundreds of chapters and tens of thousands of members across the continent; many of the vanguard of our generation were members. Although most of us never understood such rhetoric to mean an overthrow of existing institution, we did believe that we were a generation committed to change. We wanted to build a better world.

If we wanted to lead this country in new directions, our political record over the last twenty years does not indicate it. Recent history shows little sign of our influence. We have been avoiding politics. "It is the *rejection* of mainstream politics," Carpini writes in *Stability and Change in American Politics* (1986), "rather than the development of an alternative political direction that most clearly distinguishes the sixties generation." Despite our education, he reports that we are "remarkably less likely to follow politics." Indeed, had it not been for our generation, and had previous trends continued, the last three decades would have been high points in political support, involvement, and participation. We have dropped out and rejected the practices of a political tradition of which we are a part. "What distinguishes this generation," states Carpini, "is what it does not like or does not do, not what it likes or does."

Carpini based his conclusions on an analysis of the National Election Studies conducted from 1952 to 1980 by the Inter-University Consortium for Political and Social Research at the University of Michigan. The results are disturbing. Our generation is much less interested and involved in politics than are our elders. We follow politics less in newspapers, magazines, radio, and television. We register less, vote less, attend fewer rallies, belong to fewer political organizations, work less for candidates, and contribute less money to campaigns. And finally, we have contributed little to changes in the national agenda. In addition, when we vote on candidates and on issues, the bases of our decisions are unclear. Previous generations based their political decisions on party, ethnic politics, ideology, or issues. We tend to ignore such factors without replacing them with anything identifiable. For example, we are abandoning party politics. We are alienated from the two-hundred-year-old party system in general, increasingly disaffected from the Democrats and not joining the Republicans. "It is a generation without a party, and uncertain that it wants one," writes Carpini. We are outside the political mainstream where important decisions are made. This generation, Carpini documents, "rejects the norms and institutions that are central to the political system of which they are a part."

Symptomatic of our political inertia is an absence of leaders. Not only do we have no leaders with the stature of Martin Luther King, Jr., or the Kennedy brothers, but we have few leaders at all. As late as 1986, *Time* complained that our generation's "leaders are disconcertingly hard to identify." Nearly all of the leaders in the sixties came from the generation born before the 1945–64

boom (see chart p. 31). Both in politics and in music, the figures
who led us then were an older group who had not been subject to
the same influences as we had been. Perhaps this earlier generation's memories of the national unity of the war years and of a
prenuclear world help explain their dissatisfaction with American
life in their early twenties. The problem persists into the eighties.
An acclaimed work in 1987, *The Last Intellectuals* by Russell
Jacoby, laments the absence of a younger generation of writer-
thinkers in our public life such as H. L. Mencken, John Kenneth
Galbraith, William F. Buckley, Jr., and Daniel Bell. There are
hardly any intellectuals under forty-five who are known to a wide
public and who can explain the world to the general public. Where
are our true leaders? Some suggest that we may never have any.
Patrick Caddell, the political consultant, believes that this "generation may get outmaneuvered by an alliance between those ahead
of it and those behind it." We may never attain political power.

Our activism in drugs and sex is a notable exception to our
general apathy. Because of our leadership, the nation changed its
attitudes toward drugs and sex decisively. Before the 1960's, non-
medicinal drug use was taboo in the middle class; prior generations
believed that recreational drug use was evil. The use of drugs such
as marijuana and cocaine was confined to the fringes of society.
Sexual practices also differed in earlier years, because social mores
discouraged premarital sex. But we changed attitudes that had
been passed down for many generations. Amid great publicity, we
smoked marijuana, swallowed LSD, and scorned the sexual morality of our parents as a relic of a repressive past.

We cannot escape responsibility for making drugs part of
American life. In the 1960's, to be a member of our generation
meant to take drugs. We regarded smoking grass and dropping
acid as political acts that demonstrated our courage to break with
our elders, grow spiritually, and build a better society. We ridi-
culed anyone who did not use drugs, and branded them with abu-

sive adjectives such as "uptight" and "straight." Nor were we ever reluctant to urge others to use drugs. The vast majority of us assented and participated. Never did we hesitate to trumpet to the media how drugs assisted our growth. Everywhere prominent and intelligent people used drugs; and those few who abstained did not publicly object to their use. During their most popular period, drugs permeated our music, our clothes, our art, and our lives. Tens of millions of us decided that our parents and our ancestors were wrong. We began to use drugs ourselves and encouraged others to do likewise. We made drugs normal and ordinary. "There is no doubt," declared Avram Goldstein, a noted Stanford University psychopharmacologist, "that use of addictive substances is almost impossible to prevent entirely, but when a society deems something to be harmful or destructive, its use can be greatly reduced." We swept away social preventatives to drugs.

The consequences of our action lie all around us. Drugs have penetrated deeply into our society; they have even appeared in elementary schools. In the 1970's, cocaine—which is known to have severe emotional and physical effects—replaced marijuana as the drug of choice, and by the early 1980's cocaine users could be found in nearly every segment of society. Use among youth and elites was especially high. Stories of drug abuse by sports and entertainment figures—highly influential role models in this country—became commonplace in the media. In some New York social circles, it became fashionable to offer cocaine or a bright array of pep pills as *hors d'oeuvres* at cocktail parties. Another study showed that drug use was highest in families earning fifty thousand dollars or more per year. Drug abusers were most likely to be influential people such as executives, lawyers, and surgeons. The young were also especially vulnerable. Those in their late twenties and thirties were the fastest-growing proportion of users, and drug use among high school students became alarming. By their senior year, nearly two thirds of them had tried illicit drugs, and more than a third had used drugs other than marijuana. By the late seventies, more than ten percent of all high school students were using cocaine; in the late eighties that figure still exceeded ten percent. Teenagers were the only age group in the population whose life expectancy was not increasing, and some authorities believed that the cause was higher drug and alcohol use.

By 1987, AIDS, a disease transmitted both through sexual contact and intravenous drug use, had infected nearly a million people in this country. Medical authorities then forecast that the number of infected would climb to 2.5 million by early 1991, and they believed that only a minority of those infected would remain healthy. If these forecasts become true, AIDS patients in 1992 will occupy one in every ten hospital beds in New York, Los Angeles, San Francisco, and other major urban centers. Medical expenses will rise by billions of dollars. Moreover, these figures are based on the assumption that AIDS remains mainly in the homosexual and intravenous drug-user population and that its transmission rate among heterosexuals is low. Available evidence suggests that these assumptions may be correct, although the rate of incidence in central Africa is one hundred times that of the United States in all classes. At present, there is no cure for AIDS. It is fatal.

Our generation bears no direct responsibility for the existence of this deadly plague, but the sexual freedom and recreational drug use we promoted have played a large part in its spread. It is hard to imagine a disease such as AIDS breaking out in an era governed by the morality of the 1950's. We started the sexual revolution that led to a huge increase in intimate activity. Despising the sexual morality of past generations, we cast aside all inhibitions. We felt freer to engage in sexual activities. Just as with drugs, our sexual practices were a badge that we wore to differentiate us from past generations. We were the new generation with the great promise, and our first contribution to the new world that we were building would be freer sexual relations. We preached a new law, and we obeyed it. And now, we are beginning to reap the consequences.

BIRTH YEARS OF LEADERS IN THE 1960'S

POLITICS	MUSIC	OTHER
Dave Dellinger '15	Pete Seeger '19	Timothy Leary '20
Daniel Berrigan '21	Peter, Paul & Mary '38,	Staughton Lynd '29
Philip Berrigan '23	'37, '37	Tom Wolfe '31
Eldridge Cleaver '35	Judy Collins '39	Ralph Nader '34
Bobby Seale '36	John Lennon '40	Ken Kesey '35
Jane Fonda '37	Phil Ochs '40	Abbie Hoffman '36
Tom Hayden '39	Ringo Starr '40	Jerry Rubin '38
Julian Bond '40	Joan Baez '41	
Stokley Carmichael '41	Bob Dylan '41	
Huey Newton '42	Arlo Guthrie '42	
Bernadine Dohrn '42	Paul McCartney '42	
Marion Savio '43	George Harrison '43	
H. Rap Brown '43	Mick Jagger '43	
Angela Davis '44	Janis Joplin '43	
David Harris '46	Jim Morrison '43	
Mark Rudd '48	Jimi Hendrix '45	
	Donovan '46	

Where are our true leaders? Even in the sixties, we had few spokesmen who were born in the postwar years 1946–64. With few exceptions, the public figures of that time were born before 1945. Perhaps this earlier generation's memories of the national unity of the war years and of a prenuclear world help explain their dissatisfaction with American life in the 1960's.

THE DANGERS

The dangers before us are frightening. We who have begun our lives in comfort, prosperity, and peace face the prospect of ending it in far different circumstances. Nuclear war, overpopulation, and environmental disorders threaten our planet with catastrophe, while doubts grow concerning the nation's support system for the old. The doomsday clock that has appeared in each issue of the *Bulletin of the Atomic Scientists* since 1947 symbolizes our predicament. It now stands at a fraction before midnight. Ironically, our lives may become mirror images of our parents': Having spent their first years in the destitution of the Great Depression, they are spending their last ones traveling the world and living in modern homes. We, on the other hand, were born affluent and may die poor.

Nuclear War

Without doubt, nuclear warfare is our most pressing dilemma. Throughout the country, across generations, social classes, and cultural groups, a deep fear exists that our days of peace and prosperity are numbered. In *The Fate of the Earth*, a widely influential book on nuclear holocaust, Jonathan Schell discusses the extinction of mankind with a cold and persuasive realism. Only insects and grass can survive a nuclear war, he writes. Even the detonation of a single atomic bomb in a populated area will scar the planet for years and encourage the use of more atomic weapons. Once a taboo has been broken, it becomes easier to violate again.

At peace conferences, speakers often give a simple demonstration illustrating the power of the atomic weapons in American and Soviet arsenals. While the audience closes its eyes, the speaker drops a single ball bearing into a large tin can. As the noise resounds briefly in the conference room, he announces that that one ball bearing represents all the firepower expended in World War II. He drops a few more ball bearings into the can and explains that they represent the firepower of the first atomic bombs dropped on Hiroshima and Nagasaki. Then he states that he will drop into the tin can the ball bearings representing the firepower of all the atomic bombs in the world today. As the audience listens silently, the ball bearings are dropped into the tin can in such a mass that the individual balls are not heard. And they continue dropping for a time that seems endless. The audience now understands how the explosion of only one percent of the world's atomic bombs will destroy all life. A single American *Trident* submarine carries a total destructive power roughly equivalent to twenty-three thousand Hiroshimas. The United States had deployed nearly three hundred *Trident* subs by mid-1984, and an equal number of *Poseidon* subs with more than sixty percent of the *Trident*'s destructive power. The size of the Soviet Fleet is similar.

The danger of nuclear catastrophe is great because it can occur in any one of at least four ways: war between the superpowers; proliferation of nuclear weaponry among the smaller nations; terrorism; and technological error. These possibilities are listed in the order in which they are controllable.

As difficult as the task of regulating the belligerence between the USSR and the United States is, it is simpler than the complex system necessary to monitor the nuclear capabilities of the dozen countries that are becoming members of the nuclear club in the 1980's. The feuds among these nations are more volatile, the technology of their nuclear systems is less sophisticated, and the possibilities for war more numerous than they are among the superpowers. In our open and mobile societies, terrorism is even more difficult to prevent as nuclear bombs gradually become smaller and more transportable. Nuclear terrorism could occur at any time and at any place, once nuclear devices become small enough. Finally, despite the fail-safe mechanisms of our electronic nuclear-control systems, errors and malfunctions do occur. As long as machines are designed by humans and have human operators, mistakes will take place.

We are inexorably drawn to the conclusion that this generation or the next few ones may be the last to enjoy the riches of the world. Once the mass killing of a nuclear holocaust has begun, it is hard to imagine what force could hold the world back from all-out destruction. We will die either because of the blast, the thermal pulse, or the lingering radiation. Fallout will also kill the rest of life. Mammals, birds, and trees will be among the first to die. Some of the most vicious species of insects will survive, because they have very high tolerances for radiation. The remnants of humanity not destroyed by the blast, the heat, or the radiation will not survive the prolonged drop in global temperatures. Scientists are convinced that a deadly "nuclear winter" will result from the huge dust clouds a nuclear war will throw into the atmosphere. Mankind will become extinct.

Overpopulation

The rising tide of humanity also presents us with problems that appear insoluble. No matter what we do, world population will rise in the next century to a level about twice that of today. Demographers agree that the number of people on the planet will stabilize at 8–12 billion sometime between the years 2020 and 2050. Famine, wars, plagues, and improved contraceptives can only slow the growth. They cannot stop it. One measure of growth is the number of megacities, i.e., cities with populations above 10 million. In 1984, there were only three such cities in the world. By 2000, that number will increase sevenfold, and they will all be in the Third World. A recent forecast predicts that by the year 2025, 135 cities will have populations greater than 4 million. In 1985, only 42 cities were that size.

As bleak as life is today for most of humanity, conditions will become much worse in the next half-century. Year after year, decade after decade, the new billions will be packed into regions that can least afford additional consumers. Estimated growth for North America, the USSR, and Europe *combined* is less than that for either Nigeria or Bangladesh *alone*. From our comfortable compounds in the West, we will watch the numbers of malnourished children and emaciated adults increase before our eyes. Famines as catastrophic as that in Ethiopia in the mid-1980's will become as common as the annual monsoon. Just as the world of our grandparents was changed irrevocably by new technology such as the automobile, the airplane, and television, so our world will be changed irrevocably by the new billions of people who will be our neighbors on Earth. Robert Mc-Namara, secretary of defense in the Kennedy administration and subsequently president of the World Bank, has coined a new term for these people. They are the "absolute poor." Absolute poverty is "a condition of life so characterized by malnutrition, illiteracy and disease as to be beneath any reasonable definition of human decency." The World Bank has estimated that in 1980 there were 780 million absolute poor. Demographers expect their numbers to rise faster than total population.

Life in Third World countries will become nightmarish. Unemployment will worsen. Between 1980 and 2000, the world economy will add 650 million workers between the ages of twenty and forty. All but 3 percent of them will be in developing countries already hard-pressed to employ their masses. Per-capita food supplies will fall steadily. Africa is now importing food for about 25 percent of its population, although it was essentially self-sufficient in food as recently as 1970. The degradation of the environment will accelerate and affect food supplies. For example, rising demand for firewood can cause deforestation, which will cause erosion and lower soil productivity. The levels of conflict and violence will rise with overcrowding. Instances such as the war caused by a soccer game between Honduras and El Salvador will become more frequent. In 1969, a war was started between these two countries after a soccer match because Salvadorans, living in one of the most densely populated countries in the Western Hemisphere, had been illegally moving en masse into uncrowded Honduras to farm. Both sides were fighting for open land.

Authoritarianism is becoming the natural response of Third World governments to these overcrowded conditions. The dramatic decline in birthrates recently in the highly centralized societies of China and Cuba has not gone unnoticed. We should expect other nations to try policies of forced sterilization and tighter restrictions on movement. In China in recent years, neighborhood committees wielding strong social and economic powers have enforced birth-control standards strictly. Confronted with masses of illiterate, starving, and desperate people, governments of Third World countries will not be able to pay much attention to human rights and democratic ideals. Rulers will have few means other than armed force to maintain public order.

The Environment

Compared to the sudden horror of nuclear war and the progressive deadliness of overpopulation, ecological dangers are even more insidious. They are not the result of the steady accumulation of one factor—such as thermonuclear warheads or people—but are instead caused by several different environmental pathologies. We cannot identify any single problem as the unmistakable cause of future catastrophe. After much study, however, one trend is clear: The potential for global disaster is increasing.

The mildest forms of environmental pathology make the basic necessities of life scarcer and more expensive. The erosion of topsoil, the loss of cropland, desertification, the depletion of fishing stocks, the water deficit, and the reduction of mineral wealth have these effects. Topsoil loss in 1985 alone exceeded 25 billion tons, and the estimates of annual losses continue to grow. Between 1980 and 2000, total cropland growth will be only about 4 percent, while world population during the same period will grow an estimated 40 percent. The conversion of arable land to deserts is also reducing croplands. A recent UN study concluded that desertification threatens 35 percent of the earth's land surface—affecting

850 million people—and moderate desertification already affects another 470 million people, 10 percent of the world's population. Increased fertilization and irrigation may enhance the productivity of the remaining land, but farmers have used these techniques widely in the last two decades, and their returns are diminishing. The approaching water deficit also adds to the stress on world food supplies. Given existing climactic conditions and population projections, per-capita water supplies will have declined 24 percent by the year 2000. The oceans are also declining as a source of food. The oceanic catch of economically useful fish is already close to the sustainable maximum. In fact, many major fisheries are overfished. By the early 1980's, eleven major oceanic fisheries had been depleted to the point of collapse. Of the nineteen principal fisheries in the northwest Atlantic, four were depleted, nine were heavily overfished, three were moderately exploited, and the status of the other three was unknown. The story of the world's mineral stocks is similar. The gargantuan appetite of global industrial civilization is rapidly depleting our supplies. According to a 1984 analysis, known reserves of gold, silver, tungsten, nickel, aluminum, and titanium will be entirely exhausted before 2100.

The source of the so-called "greenhouse effect" is an example of the second kind of ecological pathology. These phenomena do not merely constrict supplies of food and water, but directly threaten life by major environmental change. Atmospheric carbon dioxide (CO_2) traps the heat from the sun's rays in the atmosphere. As the level of carbon dioxide in the atmosphere rises, large-scale changes in the weather can wreak havoc, inundating coastal cities by melting the polar ice caps and creating dust bowls of fertile plains. If the oceans rise the expected five to seven meters in the next twenty-five years, countries such as Bangladesh—where today 13 million people live less than three meters above sea level—will be devastated. The United States will also be affected. Rainfall patterns will shift and ruin most of the farming in the Midwest.

Acid rain also threatens to change our environment. In the
United States, acid rain is no longer confined to the Atlantic Seaboard but is becoming a national concern due to petroleum refin-
eries, power plants, smelters, and automobile emissions in the
western states. The effects of acid rain on human health are not
certain, although the National Institute of Environmental Health
Sciences has concluded that airborne acids do produce respiratory
ailments and other health problems.

Undetectable environmental problems may pose a danger
greater than either carbon dioxide or acid rain. For example, freon,
the chemical once used in aerosal cans, exposes the earth to great
dangers. In the 1970's, scientists accidentally discovered that it
eventually reached the earth's high-altitude ozone layer after emis-
sion and chemically broke down the ozone. This effect is very
dangerous, because the ozone acts as a protective shield by block-
ing most of the sun's deadly ultraviolet radiation. By 1979, U.S.
regulatory agencies had moved to phase out freon from aerosals.
But if scientists had not detected the danger of freon, an ordinary
consumer gimmick would have caused many needless deaths. We
may not be so fortunate the next time.

Our planet is not only threatened by scarcer necessities and
catastrophic environmental change but also by large-scale industrial disasters. The names Love Canal, Three Mile Island, Bhopal,
and Chernobyl have become infamous in the last ten years. The
probability of the occurrence of such accidents will not diminish in
the remaining years of the lives of this generation. In fact, their
frequency may increase. Over 35 million tons of toxic waste continue to be improperly disposed of annually. Investigators often
find drums of unknown chemicals in abandoned warehouses,
stored in small lots in random sections of cities, stashed under
elevated roadways, dumped in open fields, or poured into the
ground. Moreover, new technologies are arising in fields such as
recombinant DNA, creating fresh dangers as venture-capital in-

vestors pour millions into biotechnology firms that could unleash harmful microorganisms into the environment. And past dangers may reappear; the current unpopularity of nuclear power may not last if energy shortages return in the coming decade, as some experts predict. Finally, lower labor costs have encouraged multinational industrial firms to build new plants and relocate old ones in Third World countries where safety and managerial standards are much lower than they are in the West. Our world will continue to become more complex and interdependent as we concentrate our resources to satisfy our needs. Such concentrations may produce enormous benefits, but they also increase the possibility of catastrophes.

Our Old Age

The prosperity of our retiring parents should not delude us into thinking that our own old age will be equally golden. It will not. The economic structure that supports our fathers and mothers' comfortable retirement may collapse long before most of us are sixty-five. Their affluence is based on demographic and economic conditions that are swiftly changing and will soon be gone. The nation ages; the ratio of workers to dependents falls; and our social-security system heads toward bankruptcy. In twenty years, we will have a graying population with fewer workers and a social-security system approaching a crisis. And the problem will worsen as time passes, because the American economy itself has serious problems.

Over the next half-century, our great numbers will once again dominate this country's social processes. As we age, the nation will become older and older. In the late 1980's, about 12 percent of the population is over sixty-five. But since the elderly are already growing at twice the rate of the general population, that percentage will rise steadily. Sometime after 1990, for the first time in our history, more Americans will be over fifty-five than in elementary and secondary schools. The real explosion will occur between 2010 and 2030: the number of sixty-five-year-olds in the population will grow by one million every year during that period. By 2015, the age composition of the entire country will be as Florida's is now. And by 2040, there will be more eighty-year-olds than there are sixty-five-year-olds today. By the time the last member of our generation dies in the 2050's, the percentage of elderly in the population will have exceeded 20 percent.

Because the population is aging, the coming years will bring one of the most stunning demographic transformations in American history. The number of elderly will grow and the number of workers will decline, making the United States a nation of dependents. In the next fifty years—depending on future fertility and longevity—the working population may increase by about 6 million, while the number of elderly may jump 46 million. Between 2010 and 2025, when most of us are retiring, the number of workers may decline about 12 million, while the elderly may increase by 23 million. The ratio between workers and beneficiaries of social security illustrates the change. Today, about 3.3 workers support each beneficiary. But by 2010, that ratio will decline to 2.9; and by 2020, it will be 2.3. By 2030, fewer than 2.0 workers will support each beneficiary of social security.

These disquieting events are occurring within a troubled economy. The infrastructure of our factories, houses, bridges, and laboratories is crumbling because our anemic savings rate does not produce the capital for adequate investment. Net private domestic investment from 1980–86 was the weakest in postwar history, while the federal government continues to expand. Government spending in 1986 was 23.8 percent of the GNP, up from 20.5 percent in 1979. The nation's overall deficit expands also. In just a few years, we have transformed ourselves from the world's largest creditor into the world's largest debtor. Of course, productivity—the mother lode of the economy—is growing only a fraction as fast as it grew in the 1950's and 1960's. But the most disturbing phenomenon of all may be the hyperinflation of health costs. In the late 1980's, health-care spending was over 10 percent of GNP and has been rising at a rate of more than double the Consumer Price Index for several years. And since the aging of the population is just beginning, medical costs have no cause to stop increasing. According to one expert, Peter G. Peterson, "We face a future of economic choices that are far less pleasant than any set of choices we have confronted in living memory."

If present trends continue, we are headed for a disastrous retirement. Future workers will revolt before paying the social-security taxes necessary to support us in our old age. If we are to receive the same percentage of our preretirement income as today's retirees, future workers may have to pay as much as 40 percent of their paychecks to social security. The current rate is about 13 percent—and that seems too much. In the mid-1980's, the government issued forecasts for the social-security system using both optimistic and pessimistic assumptions. The system breaks down in several cases, although even its pessimistic model assumed an economy performing better in the future than it actually has in the past. For example, the intermediate forecast projects a deficit for Medicare in 1993 and a depletion of its reserves by 1998. But if the reserves of the disability fund and the pension fund are used to support Medicare, the entire system will become bankrupt long before we need it. "The assumption that each working generation will take care of the one that preceded it is finished," declares Senator David F. Durenberger, a national leader in this field.

Few generations have received as many gifts, failed in as many ways, and faced as many disheartening dangers as do we, the gifted generation born in the years following World War II. We grew up in a profound peace and enjoyed personal liberties and advanced technologies envied by our contemporaries around the world. We were confident in the future, optimistic about our prospects, and proud of ourselves. But as we near the midpoint of our lives, we see that our circumstances have changed. Not only do we have trouble matching our parents' standard of living, but we have also destroyed families and abandoned children in a quest for the fulfillment of a promise that has yet to materialize. Moreover, we have unleashed upon the middle class the scourge of cheap drugs and made it vulnerable to a sexually transmitted disease as deadly as medieval plagues. And our future no longer looks bright. The nuclearization of the world continues inexorably, while global population increases unchecked, and environmental disorders erupt in the soil, water, and air. The future is frightening. What shall we do?

The answer to this question lies in our heritage and in our future. We have not been born isolated and alone but are the children of a rich civilization. Much has been bequeathed us. In addition, we have in our hands the power to initiate a new and brighter era for humanity. A society centered on Earth and extending the breadth of the solar system will help resolve many of our ills. Enlarging the human domain by opening the space frontier gives hope to us all.

CHAPTER TWO

Manifest Destiny and the Space Frontier

Gaston's San Francisco Ice Cream Parlor in Djakarta, Indonesia, is an authentic example of America. It was designed in California, it has a San Francisco Victorian touch, and it receives all the ingredients of its products from the United States. Gaston's is very popular among Indonesians. More than four thousand people jammed into it when it opened in 1983; and eighteen months later, the company still served as many as three thousand customers per day and had plans to open a second Djakarta outlet. Gaston's is not the only sign of American influence in Southeast Asia. Swenson's, another California ice creamery, is down the street; and the San Francisco Ice Cream Company is located across town. Young Indonesians dressed in Esprit sweatshirts, tailored Levi's, and jogging shoes throng these shops in the evening.

The homage Asians pay to the icons of a foreign culture illustrates a worldwide phenomenon that is centuries old. Since Christopher Columbus discovered the Americas in the late 1400's, Western civilization has invaded and indelibly marked every culture on Earth. Its dress, language, religions, and technology appear everywhere on the planet. In the early decades, the West planted colonies in sparsely populated regions, e.g., the English in North America, the French in Canada, and the Spanish in South America. In a later phase, the West conquered and ruled other civilizations, as in India, Africa, and China. By the twentieth century, colonization and conquest had become outmoded methods of imperialism, because cultural colonization had replaced them. Western civilization no longer had to force its ways on the unwilling. Spurred by the captivating images of television and cinema, hundreds of millions learned to despise their native customs and to regard Western life as "modern."

Western civilization has expanded more rapidly and more extensively than any civilization in human history. Most of the other twenty-odd civilizations that have arisen are superior to the West in some way. Ancient Egypt may have been more stable, India more spiritual, and Africa more natural. The Chinese perhaps governed better; the Romans waged war better; and Islam was less commercial. But none of these cultures grew as widely and as quickly as the West. Western civilization excels in exploration, settlement, and development. When it began its unusual career in Europe about A.D. 1500, humanity was divided among several civilizations having little connection with one another. For generations thereafter, this rough and often graceless culture sent out explorers, merchants, soldiers, and emigrants to develop its Great Frontier—the Americas, Africa, Australia, and the Pacific Islands. By the mid-19th century, the West had triumphed everywhere; by the early twentieth century, the historian Arnold Toynbee could announce, perhaps prematurely, that "of the living civilizations every one has already broken down and is in the process of disintegration except our own."

Called by the historian William McNeill the "most thoroughgoing translation of European-type society," America is the principal heir of Western civilization. Part of the rich heritage it has received from the West is a collection of skills in expansion that cannot be matched by any other nation. This fact should not surprise us. Five hundred years ago, North America was a desolate wilderness; today, it contains a highly urbanized society populated by immigrants and the descendants of immigrants. Unlike Rome, unlike China, unlike Great Britain, we never learned how to weld far-flung peoples into a unified empire. Instead, we learned how to build prosperous and free communities in places once beyond the frontier of civilized life.

We call the traditional ideology of American expansion Manifest Destiny. The tendency of our people to move outward beyond our borders began long before it was named. Movement west had begun almost as soon as the first colonies were planted in the seventeenth century; and prominent members of the founding generation envisioned an America much larger than the original thirteen colonies. The notion that this movement had an inevitable consequence—the settlement of a certain portion of the Western Hemisphere—gradually came to be accepted as a national destination that was plain to all, even if citizens and politicians disagreed among themselves about the proper extent of the expansion.

Although the idea of destiny has often been abused, it has served this country well. The vision of a free nation stretching from the Atlantic to the Pacific inspired generations of men and women around the world. In the United States, it focused the energies of politicians and gave direction to the natural inclination of the people. Now, the idea of destiny must be updated. We stand on the verge of a new phase of our national life, perhaps our greatest. The global crises of the environment, overpopulation, and nuclear war, our position as chief heir of a great expansionary civilization, and the world's need for our extraordinary skills propel us toward space. The obvious destination of the United States is the exploration and settlement of the solar system. American movement into space continues centuries-old trends of expansion; it strengthens fundamental values of our civilization; and it gives the world a kind of leadership that no other nation can provide. Failure to rise to our Manifest Destiny endangers ourselves, our species, and our planet. If humanity is to build an extraterrestrial civilization and evade the crises plaguing its home planet, America must devote its talents to expansion into the solar plain, which extends from Mercury, past the inner planets, through the asteroids, and beyond the outer giants to Pluto.

PRINCIPAL HEIR OF AN EXPANDING CIVILIZATION

The eleven-volume *Story of Civilization* by Will and Ariel Durant is one of the most ubiquitous histories of our time. Its multicolored covers often grace book-club advertisements and appear in many private libraries. Laboring over a period of nearly fifty years, the Durants wanted to write a history that was a biography of mankind accessible to the average reader. By many measures, they succeeded. *The Story of Civilization* reflects their painstaking research and their talent for entertaining storytelling. Each volume was a best-seller, and their penultimate work, *Rousseau and Revolution*, received a Pulitzer Prize in 1968.

Their work illustrates a common misconception about world history. The Durants mistitled *The Story of Civilization* in a grossly ethnocentric manner. The history that Will and Ariel Durant wrote is only the story of *Western* civilization. Although part of the first volume covers the civilizations of China, India, and the Middle East, the remaining ten volumes are devoted exclusively to the story of European life. The Durants' history promotes the widespread belief that Western civilization forms the substance of mankind's history. This opinion is profoundly mistaken. As important as the West is, the human story on Earth is much broader and richer than the contributions of a single civilization.

What is a civilization? We are social animals, and civilization has arisen out of this disposition of humanity to live in groups. (In fact, society antedates our species, since our origin almost certainly lies with mammals who were themselves social creatures.) Civilization is one kind of human group; primitive society is another. Primitive societies are short-lived, involve a small number of people, and cover a narrow geographical area. They vastly outnumber civilizations. (Arnold Toynbee characterizes them as rabbits.) Civilizations, on the other hand, are massive societies. Their populations number in the millions, their territories range over thousands of square miles, and they endure for periods much longer than the span of a human life. (Toynbee refers to them as elephants.) Civilizations also create cities. A civilized man is a man who lives in a city; the very term "civilization" derives from a Latin word signifying an inhabitant of a city. Civilizations are distinguishable from primitive societies by their cities, population, territory, and longevity.

Humanity has lived in civilizations for a relatively brief period of time. Our biological age is far greater than our civilized age. The earliest generally accepted representatives of the genus *Homo, Homo habilis*, walked upright and used tools about two million years ago. About 500,000 years ago, a more advanced species, *Homo erectus*, used fire, possessed more refined tools, and lived in caves. Our own species, *Homo sapiens*, emerged about 100,000 years ago and soon covered the planet. These first men and women were no different from us. They had our physical powers, our intelligence, and our emotional range. For tens of thousands of years, our people roamed the land in tribes, hunting animals and searching for edible plants. This was our first age, an era that the biblical account of Eden recalled. In the second age, we lived in small agricultural communities, which first developed ten thousand years ago in the Middle East. This period of primitive agricultural communities lasted about four thousand years. Civilized life is the third of mankind's great ages. The first civilizations arose no more than six thousand years ago. The period of time in which humanity has lived in civilized societies may seem immense, but it is only an instant compared to the geological ages of the planet and the evolutionary age of our line.

During this third age of about six thousand years, our species developed seven civilized traditions within which historians have identified more than twenty individual civilizations. Four of these traditions originated in the so-called Old World (Asia, Europe, and Africa) and three in the New World (North and South America). Each tradition grew independently of the others for centuries, although traders often introduced innovations from one civilization into another. Each tradition was also dynamic. Especially in the Old World, civilizations rose, declined, and gave way to a successor. For example, Islamic civilization, which began about A.D. 700, rose in the region of the Sumero-Akkadian civilization, which had flourished three thousand years earlier.

The relationship among the civilized centers of humanity has not stayed constant. The three civilizations of the New World—the Aztec, Mayan, and Incan—developed independently until the fifteenth century A.D., when they were overwhelmed and destroyed by the more advanced technology and more powerful military of the Old World. They were much younger and therefore less advanced than their counterparts in Eurasia. For two thousand years, the civilizations of the Old World—China, India, Europe, and the Middle East—coexisted in rough equilibrium. From 500 B.C. to A.D. 1500, they had little connection with one another and developed independently. Each civilization worshiped different gods, obeyed different rulers, and practiced different customs. China had its decorum and its dynasties, India its transcendentalism and its castes, Europe its territorial states, and the Middle East its bureaucratic empire. Few suspected the full extent of human life, and those who did were likely to have ideas more fanciful than accurate.

Several times during those two millenia, one of the four Eurasian civilizations upset the balance by expanding into regions traditionally ruled by its neighbor. But none succeeded in permanently altering the composition of the world's cultures. First, Alexander the Great threatened to obliterate the Middle Eastern civilization in the late fourth century B.C. Next, Indian civilization under the Gupta dynasty (c. A.D. 320–535) spread Buddhism throughout Central Asia, China, and Japan. Finally, Islam expanded steadily at the expense of both Europe and India after Muhammad died in the seventh century. In each of these uprisings, the expanding civilization attempted to establish a wider cultural supremacy at the expense of its neighbors. But all failed until Europe began its expansion in the fifteenth century. The West is the first civilization in history that has succeeded in imposing its culture on the entire world. All previous attempts failed.

An intelligent and informed observer in the fifteenth century would not have suspected that the semibarbaric nation-states of Europe would succeed in altering the two thousand-year-old balance of power among the world's civilizations. Europe was poor, backward, and feudal. It was inferior in culture to the Ming dynasty ruling China, and in military force to Islam. European art and culture did not approach the delicate perfection long associated with Asia, and its modest expansion in the previous few centuries seemed feeble compared to Moslem triumphs. (Islamic civilization reached one of its peaks when it captured Constantinople in 1453.) In truth, Europe in the fifteenth century surpassed other civilizations in few ways, and no one could have foreseen its astounding progress in the next five hundred years.

The growth of Western civilization in the last five centuries is unprecedented in history. The West in the late fifteenth century was confined to the small peninsula of Europe, an outcropping of the greater land mass of Asia. But within four hundred years, it mastered the planet. By 1600, the Spanish and French had bases in North America; the Portuguese had settled in Brazil and set up trading posts around Africa and India; and the Spanish had claimed the Philippines. By the early 1800's, the West had established itself on every continent, and by the end of the century it had opened Japan, divided China, and parceled out Africa. No one could deny that the West dominated Earth. The millennial balance among the Old World civilizations—which had begun about 500 B.C.—had broken down, and most of the planet had become satellites of Europe. "From the perspective of the mid-twentieth century," wrote the historian McNeill in 1963, "the career of Western Civilization since 1500 appears as a vast explosion, far greater than any comparable phenomenon of the past both in geographic range and in social depth." The expansion of Europe destroyed the equilibrium that had existed for two thousand years among the world's civilizations, and it established Western civilization as the dominant culture of the planet.

Of the twenty-one civilizations that have flourished since the first representatives of this kind of human society appeared, four in addition to Western civilization have endured into the late twentieth century. The civilizations of Islam, China, India, and Russia still exist, although their vitality as cultures independent of the West may be fading. The regime that has ruled China for four decades rejects its past as a matter of state policy. India derives its government from a European source and possesses both nuclear and space technology. Russia—never far removed from the West in the first place—represses its traditional religion in favor of science. Islamic civilization may be more vigorous. However, in the furiously anti-Western, intensely Islamic Iran in the mid-1980's, quotations from Voltaire, Samuel Johnson, and John Keats were printed alongside those of great Muslim philosophers in the *Tehran Times*. And despite the austere discipline of their clerical rulers, Iranians display a keen appetite for popular American music, television, and film. At least one historian believes that of existing civilizations only the West will survive. In the twentieth century, wrote Arnold Toynbee, Western civilization is "apparently in the singular position of being the only one . . . whose present state and future prospects might still be open questions." No wonder the Durants succumbed to what Toynbee has called "the distorting, egocentric illusion" that Western civilization *is* civilization.

We can define Western civilization by its duration, its relations to other civilizations, and its territory. The West has existed for about one quarter of the time humanity has been civilized and living in civilizations. Established about fifteen hundred years ago, when the Western region of the Roman Empire was collapsing, Western civilization grew out of the Hellenic civilization of the Greeks and the Romans, which had developed from the Aegean civilization that had preceded it. The territory that the West occupies is a minor part of the total habitable area of the surface of this planet. It is bounded in the West by Alaska and Chile and in the East by Finland and Dalmatia. Included within these four points are North and South America, and Europe itself, which is the tip of Asia's western peninsula. Western civilization also possesses outlying footholds in South Africa, Israel, Australia, and New Zealand.

The worldwide prevalence of Western languages and religion indicate the influence the West has in the world today. Our dominant culture spread its religion and languages beyond its borders, as did so many other conquering civilizations. What religious leader has more renown than the pope, the most prominent Christian religious leader? What language is more popular than English, the most common language of the West? Christianity, the religion of the West, has more adherents than the religion of any other civilization. In 1985, more than a billion people counted themselves Christian, while about 600 million considered themselves Muslim (Islamic civilization). Buddhism and Hinduism (Indian civilization) claimed about 700 million believers; but Confucianism (Chinese civilization) had less than a quarter of a billion. A statistical study of Western languages proves they too have spread enormously. By some estimates, nearly 2 billion of the world's people speak a European language. Speakers of one of the Chinese languages number no more than a billion, while speakers of Arabic and Hindi combined do not exceed 500 million.

Western civilization is unique among the twenty-one civilizations that have arisen in the last six thousand years of this most recent phase of human life. Among these civilizations, the West has unequaled skills in exploration, settlement, and development. These abilities allowed the West to expand over the whole habitable surface of the planet and produce a world culture. In the fifteenth century, Europe faced a great frontier consisting of North America, South America, Africa, Australia, and thousands of islands in the Pacific. In the next few centuries, it sent out hundreds of explorers and adventurers, of whom Columbus, Magellan, Vasco da Gama, and James Cook are only the most famous. It also sent out millions of settlers in a freely undertaken, individually financed, long-distance migration without parallel in world history. In the seventy-five year period from 1846 to 1920 alone, more than 46 million Europeans emigrated. The result of this unprecedented expansion is a global civilization in which economic development has reached such an extreme that population is outstripping resources and the planetary environment is deteriorating.

The world civilization of the late twentieth century may already be metamorphosing into something new that is an amalgam of many cultures. If so, that new civilization will, doubtless, possess at least one peculiarly Western trait: a hunger for expansion that cannot be satisfied. Western exploration of this planet did not end until the twentieth century, when explorers and statesmen turned their attention to Antarctica, the last unexplored and unclaimed continent on Earth. Historians call the first two decades of the 1900's the heroic age of Antarctic exploration. Significantly, even at this late date, the explorers and the nations making territorial claims were Western. The British sent sled probes deep into the interior in two expeditions (1901–4 and 1907–9). And the Norwegian Roald Amundsen reached the South Pole in December 1911, a month before the British arrived led by Robert Scott. Territorial claims followed quickly. Between 1908 and 1942, seven nations—Great Britain, New Zealand, Australia, France, Norway, Chile, and Argentina—decreed sovereignty over pie-shaped sections of the continent. No representatives of any of the other four civilizations still inhabiting Earth sent explorers or made claims. Historical patterns do not change easily.

The Western drive to explore continues. The Royal Geographic Society of London, which has roots traceable to the eighteenth century, sponsored expeditions to Antarctica in the 1940's, to Mount Everest in the 1950's, and to remote parts of Brazil, Kenya, Oman, and Malaysia in the 1960's and 1970's. But the most important explorations of the twentieth century have been made by the United States, which furthered the Western tradition of expansion in a way that can be compared with the great age of Columbus and Magellan. In 1969, it landed men on the Moon for the first time. And a magnificent series of interplanetary probes launched by the United States visited all the planets between 1962 and 1989 except distant Pluto. The astronomer and planetologist Carl Sagan has rightly called this period "the golden age of planetary exploration."

His pride is well founded. Few eras in Western history can match the achievements of these times. Six American expeditions landed on the Moon between 1969 and 1972. And probes bearing the names *Mariner, Pioneer, Viking*, and *Voyager* left Earth to survey the inner planets, traverse the asteroid belt, and fly by the gas giants beyond. Mission after mission was a success. Thirteen spacecraft scouted the Moon in preparation for the *Apollo* landings. Venus was probed in 1962, Mars in 1965, Jupiter in 1973, Mercury in 1974, Saturn in 1980, Uranus in 1986, and Neptune in 1989. The most glorious moments were the two landings on Mars in 1976 by *Vikings 1* and *2*, which transmitted to Earth the unforgettable color photographs of the Martian landscape. Surely these extraordinary triumphs justify Sagan's proclaiming "the present moment a pivotal instant in the history of mankind." Surely these feats allow him to boast—in the expansive language characteristic of the great explorers of Western civilization—that "not many generations [are] given an opportunity as historically significant as this one."

The United States is the favored child of the West, the mighty culture that has won the six-thousand-year competition for supremacy among the world's civilizations. We occupy a continent that was one of the main regions of European expansion; we embody Western traits in their most extreme form; and—most significantly—our expansion from a narrow beachhead along the Atlantic Coast into a continental nation mimicked Europe's own expansion around the world. Until this century, the great tides of immigration that peopled America originated in Europe. Our land was one of the chief destinations of the settlers who were inspired to leave their ancestral homes for an unknown life. Western civilization promoted self-government, capitalism, and technology. In the United States, these values flourished prodigiously. We are famous for our passion for freedom, our desire for wealth, and our love of change. But the trait that marks us most as the offspring of the West is our emphasis on growth, from the territorial growth of earlier eras to the economic and personal growth of our own times. The process of immigration, settlement, and development took place on three continents, a large part of a fourth, and on thousands of islands. Yet it is the drama of the American West that has become part of the folklore of the world. Perhaps the identity between the popular name for the American frontier and the civilization of which it is a small part is a significant coincidence. A father bestows his own name on the child who bears his greatest hopes.

The *Apollo* Moon landing in 1969 and the brilliant series of interplanetary probes in the 1970's and 1980's fit neatly into historical patterns. If we are indeed the chief offspring of a civilization notorious for its acquisition of new territory, how appropriate it is that one of our citizens took the first steps on a new world. The fires that power Western civilization's engines of expansion have not died. When the Wright brothers began the movement into space at Kitty Hawk in 1903, they were continuing traditions of expansion that were centuries old. These traditions are not vague historical trends but reflect realities embedded deeply in the souls of Western men. The momentum that has driven Western civilization for five hundred years has not been exhausted.

The United States is the principal heir of a civilization that possesses the greatest tradition of exploration, settlement, and development that the world has ever seen. Five centuries ago, Western civilization was no more than a mediocre third world compared to the splendors of imperial China and the military successes of crusading Islam. At that time, human life was fragmented among a half-dozen cultures having little effect on one another, while large expanses of the planet lay desolate or sparsely populated. This condition, which had lasted for thousands of years, is gone forever. By the middle of the nineteenth century, the West proudly colonized the entire planet; and by the middle of the twentieth century, it had created a world civilization in its own image, stigmatizing other forms of civilized life as backward. America is the favored child of this grand civilization. We received her greatest gifts and have exploited them without quarter. The first centuries of our history showed that we too know how to expand, to settle, and to develop; and the most recent decades show that our appetite for growth is as sharp as that of our forefathers.

The United States is continuing the Western tradition of expansion by beginning the exploration and settlement of the solar system. In the fifteenth and the sixteenth centuries, Columbus discovered the Americas and Magellan encircled the globe. In the seventeenth century, La Salle sailed down the Mississippi into the heart of North America. In the eighteenth century, James Cook explored the South Pacific. In the nineteenth century, David Livingston explored sub-Saharan Africa. In the early twentieth century, Scott and Amundsen explored Antarctica. In the late twentieth century, the United States landed men on the Moon and explored most of the planets in the solar system. We are part of a vast movement of life that started centuries ago and has continued to the present day. It has not stopped.

AMERICAN SKILLS

The Rocky Mountain Fur Company is an example of the kind of venture that hastened the development of the American West. This company was the first to depend on trapping by its employees rather than trading with Indians. William Ashley, a general of the St. Louis Militia, founded Rocky Mountain in 1822, when he advertised for one hundred young men to work in the wilderness for one to three years above the headwaters of the Missouri River. Among those he hired was Jedediah Smith. Unsurpassed as an explorer, Smith has been compared to Meriwether Lewis and William Clark. Before he was killed by Comanches at age thirty-two, he discovered a gateway pass to the West in 1824; he led the first expedition to southern California from the Great Salt Lake in 1826; he became the first white man to cross the Sierra Nevada from west to east in 1827; and he led the first expedition up the California-Oregon Coast in 1828.

The Oregon Trail was one of the great highways of the old West. Starting in Independence, Missouri, it stretched two thousand miles through Indians, grizzlies, cholera, and bitter weather to Willamette Valley in Oregon. The first emigrant train carried eighty men, women, and children west in 1841. On good days, ox-drawn wagons traveled twenty-five miles; on bad days, they went no more than ten miles. In the "Great Migration" of 1843, two hundred families—about one thousand people—and hundreds of cattle made the journey. These caravans were only the beginning of a mass movement that made Oregon an American territory in 1849 and a full-fledged state ten years later.

Risky new businesses like the Rocky Mountain Fur Company, intrepid explorers like Jedediah Smith, and dangerous treks like those on the Oregon Trail are not isolated instances in American history. Americans have repeated these kinds of experiences a thousandfold since the first European colonists arrived here in the 1600's. Because of the conditions we faced, we had to learn how to expand the region of civilized life. We had to learn how to explore, settle, and develop. This harsh imperative confronted us for centuries, and the disposition it fostered has not died. By the early 1980's, hundreds of new companies had already been formed at great financial risk to commercialize space. The potential loss to careers and personal lives for the founders of these enterprises was not trivial. And our propensity to settle new lands remains. In the mid-1970's, a space society was started with the goal of disbanding itself at a mass meeting in a space colony; by 1984, it had almost ten thousand members and a national network of over seventy chapters.

As an expansionary society, and as a scion of an expansionary civilization, America has learned skills that allow it to grow by acquiring new territories. Of course, given an opportunity, any country can finance explorers, send out settlers, and develop a new economy. We will witness many attempting to do so in the next century. But they will not explore, settle, and develop as well as the United States. No other nation has distilled so purely the expansionist traditions of Western civilization, and no other nation has so schooled its people in the lessons of growth. Other nations may govern better than we, others may conquer better than we, and others may enjoy the pleasures of life better than we. But none can equal our excellence in the art of creating communities in new lands.

We have not achieved this position by chance. America has become an extreme representative of the most expansive civilization in history because the West's most distinguishing traits are overgrown within us. The West differs from other civilizations by the control it has over the natural world, the mechanization of its economy, and its tendencies toward political equality. These characteristics distinguish the West from other forms of civilized life and help explain its extraordinary ability to expand. The United States has these traits to excess. We are an outstanding example of a Western nation in our inventiveness, our devotion to business, and our love of equality. Our prowess as inventors became apparent early in the nineteenth century; few nations can boast of having invented more of the technology that makes the modern world possible. We also embraced capitalism. Until the administration of Franklin Roosevelt in 1932, American businessmen operated under one of the most radical forms of laissez-faire capitalism found anywhere. Even today, the climate for commercial individualism in the United States remains one of the most favorable in the world. Of course, America's love of freedom and equality is legendary. We are a middle-class society in which great wealth and power are concealed and rarely transferred from generation to generation. If to be Western means to be technological, commercial, and democratic, America is Western to an extreme.

We have also enjoyed advantages that other nations have not. Europe herself entered the technological, industrial, and democratic world handicapped by outmoded social and political institutions. Even today, European countries suffer from vestiges of feudalism and aristocracy, such as the status that descendants of old families retain. But Americans started fresh. We have always practiced democracy, furthered science, and invented technology without the artificial restraints of a dead past. European settlements on other continents also had disadvantages that did not burden us. Political and economic life in North America was vigorous and independent. Our ancestors were self-governing, while settlers on other continents passively submitted to imported masters ruling them through bureaucratic, oligarchic, and ecclesiastical hierarchies. Moreover, large native populations did not affect colonial America to the degree they did in Africa, Asia, and Latin America. We were free to build a new society undiluted by an underlying heritage from another civilization, and thus the society we built was almost wholly derived from European traditions. European settlements in other lands were not as pure an outgrowth of Western civilization.

America's great skill consists of the ability to create self-sufficient communities in new lands. That is the true meaning of expansion: extending the domain of civilization by creating the enclaves that nourish human life and all that attends it. The United States excels in this complex technological, economic, political, and governmental skill that we learned as we converted the wilderness of America into an advanced urbanized society. We know how to make the tools, appliances, and devices to adapt the natural world to our benefit. We know how to stimulate large groups of people to work for economic goals. We know how to rule ourselves without destroying initiative and avoiding the classic errors of colonialism. And we know how to integrate new territories into our commonwealth. Our aptitude in these areas enables us to create self-sufficient settlements and gives us preeminence as an expansionary power.

Americans know how to develop technology. A country of entrepreneurs, we have always been a practical people and have enriched life with many inventions. The success of the *Apollo* Moon program, the numerous interplanetary probes, and the space shuttle are the best contemporary evidence of our talent. Thousands of engineers working a quarter-century developed the innovations needed to land on the Moon, probe the planets, and orbit the shuttle. No other nation can match this record. The USSR has supported its space program lavishly and far more consistently than we have ours. Yet many of its space technologies are primitive in comparison, despite its clear lead in transportation systems. For example, the Soviets launch twenty-five to thirty photographic reconnaissance satellites each year that serve the same function as a handful of U.S. satellites.

The political principles that organize American life encourage an intense individualism that can lead to high achievement. American society gives individuals liberty and stimulates them to develop their powers to the fullest. In every field, the individual talent recognizes that self-cultivation can lead to distinction. The leaders among us learn that they earn their greatest rewards by forming groups of high-performing individuals who will work to solve social needs. This characteristic of our political life is not common. Few other societies permit the individualism required for self-development or the liberties that permit leaders to create free associations.

The clearest illustration of this phenomenon in the United States is in our economic life. We have always celebrated entrepreneurs, those leaders who know how to build organizations that supply the goods and services we consume. These men and women possess a high political art. They can recognize a need the country has and create an institution to fill it by mobilizing the energies of other self-devoted individuals. We honor entrepreneurs extravagantly for their leadership, giving those who succeed on the largest scale the greatest rewards. Henry Ford earned his millions of dollars because the enterprise he founded provided millions of people low-cost transportation; Steve Jobs earned his by providing low-cost computing. Like other entrepreneurs, these men knew how to solve society's problems by building an organization that employed the talents of many people.

In recent years, we have become more sophisticated in fostering this skill. We now have a system of education to develop aspiring entrepreneurs, lower-risk variants of entrepreneuring, and an institution that assists advanced entrepreneurs. A flourishing subculture with its own legends and heroes teaches the art of entrepreneurship through magazines, seminars, conventions, books, television shows, and college courses. In some parts of the country—for example, Boston's Route 128 and California's Silicon Valley—a dense support network exists to strengthen the least impulse toward business creation. Franchising is a lower-risk form of entrepreneurship, in which the franchisor sells a successful business formula to a franchisee, who accepts a lower return for lower risk. Franchises may generate half of all retail sales by the year 2000. Business incubators are another way new ventures can reduce risk. For the rent of market-rate space, they provide entrepreneurs consulting help and resources that are often too costly for new companies, such as phones, secretaries, computers, and copying machines. And for those who are advanced entrepreneurs, venture capitalists provide the finance and expertise to build $100-million-plus companies. Only in a nation devoted to expansion could such customs arise. The genius of America's political system consists of its ability to release for the pursuit of common goals the full power of the individual talent through entrepreneurship, franchising, and venture capitalism.

The countries south of the Rio Grande have failed to achieve the political and economic success of the United States. Uncontrollable inflation, pervasive poverty, chronic military rule, frequent guerrilla warfare, and political disunity plague Mexico, Central America, and South America. The United States has been able to build an economy with one of the highest living standards in the world, unify a continent under one government, and develop a society free from military oppression and political terrorism partly because of the different ways the Americas were settled. Representative governing institutions and religious toleration did not exist in Spanish colonies in the sixteenth and seventeenth centuries. All rule came from above, and diversity was discouraged. The French governed Canada in a similar way. Only Catholic colonists were permitted, and Paris gave the settlements a government that was both despotic and paternal. English colonists, on the other hand, arrived in North America with a firm belief in freedom of speech, press, and assembly. And they created self-governing communities that exhibited boundless individual initiative and in which political liberty and social mobility prevailed to an unusual degree. Economic sufficiency and self-government flourished in this environment. From such roots have sprung America's prosperity and democratic traditions.

Among all the space powers, the United States is the most likely to avoid the historic mistakes of colonialism and follow the English model of settlement. We know how to rule ourselves without destroying initiative. Space pioneers will live on dangerous and treacherous new worlds in which their judgment and personal resources may make the difference between the survival or extinction of their settlements. They must rely on themselves and not be ruled by Earth-based controllers. Nations that restrict the liberty of their citizens on Earth will not grant them greater freedom in space. American space settlements will resemble early English settlements in North America. Their disposition to individual initiative and self-rule will enable them to survive the rigors of the early years and lay the foundations for extraterrestrial nations resembling twentieth-century North America.

An unmistakable sign of America's expansive character is our government, which facilitates expansion into new lands. We have a process for converting raw wilderness into functioning and equal parts of the nation. The historians Allan Nevins and Henry Steele Commager have called the Northwest Ordinance, enacted in 1787 and tested for over a century, "a wise plan which did much to make the United States the country it is." This law set a pattern of orderly and progressive settlement of new regions leading to self-government by regular stages. In the first stage, Congress declares a new area of settlement a territory and appoints a governor and judges, who can make laws subject to Congressional veto. When the population in the new territory reaches five thousand, Congress creates a legislature with two chambers. Finally, after the population reaches sixty thousand, the territory may become a state equal to the existing states. This process established thirty-five states between 1787 and 1912 and created a wealth of legal precedents for integrating new lands into our country.

The structure of the Constitution and our experience with state government give us two other advantages in expansion off-planet. The federal structure of the Constitution is ideal for governing new territories. New settlements can concentrate on solving the complex problems of frontier life without the distractions of foreign affairs and a variety of other issues that are better handled by a central government. For its part, Washington will tend not to interfere in the local affairs of space settlements, as it tends not to interfere in state and local government now. By so doing, it will avoid the disastrous mistakes of the French and Spanish governments, whose excessive intervention in their sixteenth-century colonies proved ruinous. The success and experience this nation has had in fashioning fifty state governments is also important in the development of space communities. The tripartite division of powers, bicameral legislatures, and the town/county system of local administration established in most states define clear models for future government-makers. No other people can claim such expertise in self-rule.

For the next few decades, the most important task in the settlement of the solar system is the creation of self-sufficient communities. The survival of humanity's extraterrestrial enterprise depends on our ability to build settlements off-planet that can support and govern themselves. The alternatives are obvious. Earth will not support space cities that do not justify themselves economically; and the enormous distances of interplanetary travel require some form of self-government. When we have mastered the art of planting colonies that can pay their way and avoid anarchy, the movement of humanity off our home planet will have begun in earnest.

The talent that America has for creating self-sufficient colonies confirms our preeminence as an exploring, settling and developing power. A people that is self-sufficient is a people that knows how to govern itself without restricting its human resources; it knows how to seize every opportunity for profit; and it knows how to invent the thousand-and-one adaptations that life in a harsh environment demands.

The political traditions of self-government and democracy the first colonists brought to these shores are the same traditions that we have recreated in fifty state governments. What could be more probable than that they will continue to function in American space settlements? The extreme individualism that has always driven Americans to build new organizations and earn new wealth has intensified in our era. This same spirit of individualism will also drive our space communities to economic independence. And we have always been an inventive people. Our practical ingenuity will not fail us now.

Mankind needs a people that has the talent to transform the species into a spacefaring race; it needs a nation that has the ability to put the world on the path that leads to humanizing the solar system. We are that nation. This country has the political, economic, and technological skills needed to build self-sufficient colonies. The United States is the ideal leader of mankind's movement off this planet. We are exploratory, migratory, and entrepreneurial to a degree not matched by other nations. We have acquired superb skills in expansion that are technological, political, social, and governmental. We know how to devise new technologies, we know how to liberate the energies of masses of people, we know how to create self-sufficient settlements, and we have the governmental structures for dealing with new territories. The United States is the greatest exploring, settling, and developing power the world has ever known. We have been bred to it.

AN OBVIOUS DESTINATION

A meeting sponsored by the American Enterprise Institute in late 1983 illustrates the unusual position of America in the world. The session concerned constitution-writing, and over twenty countries sent representatives. In this convocation, the United States was conspicuous for several reasons. Except those from the United States, the delegates of every nation had played a major role in drafting their country's constitutions. Except that of the United States, no constitution was older than France's, which had been written in 1958. Most had been written in the 1970's. Again, except the United States, every country had based its constitution on models written earlier. With a constitution nearly two hundred years old, America is extraordinarily stable. Other nations have not been as fortunate. The French have had five constitutions during the same period, and within four years of this meeting Nigeria had discarded its constitution. Not only is the United States distinguished from other nations by its possession of a written constitution, but it also possesses one of unusual longevity.

Incidents such as this one feed Americans' enduring sense of destiny and mission. Whether because of our wealth, our liberty, our advanced way of life, or the character of our political traditions, we believe that we are set apart from the rest of the world. "From its earliest beginnings," wrote the historians Nevins and Commager of America, "its people have been conscious of a peculiar destiny." When John Winthrop described "the city set

upon a hill'' that he and his fellow Puritans intended to found in
the early 1600's in Salem, Massachusetts, he expressed the dream
of many future generations. Throughout our history, we have believed that we enjoyed opportunities that demanded something
more from us. We have believed that we received gifts that would
allow us to accomplish something high and rare.

What we have hoped to accomplish has not remained the
same. At one time we strived for religious liberty; at another time
for an independent government, ruling with the consent of the
governed; at another for a wider democracy including all people
regardless of property, race, or sex. Still later, we conceived our
mission to be to save the Union; and still later, we fought to keep
democracy in Europe. In recent times, we have wanted to eliminate the poverty that threatens the stability of our society and to
preserve the environment that has benefited us so greatly. At different times in our history, our sense of being special or gifted has
led us to strive for different ideals.

Our intellectual and political leaders have reflected this sen-
timent. In the nineteenth century, Herman Melville, author of
Moby Dick, wrote that ''We Americans are the peculiar, chosen
people. . . . God has predestinated [and] mankind expects great
things from our race; and great things we feel in our souls.''
Woodrow Wilson, whom historians consider one of our greatest
presidents, spoke on the same theme in 1919. ''I, for one, believe
more profoundly than in anything else human in the destiny of the
United States. I believe that she has a spiritual energy in her
which no other nation can contribute to the liberation of mankind.'' When Ronald Reagan stated that he believed that ''we are
destined to be the beacon of hope for all mankind'' in his second
inaugural address in 1985, he was echoing an idea that has a
lengthy genealogy.

One mission that has moved us is Manifest Destiny, the traditional ideology for American expansion. Manifest Destiny is one

expression of the idealism that has always been present in American life. It meant that the nation and its institutions were to spread over a larger area, an area that was never clearly defined. For some it meant expansion to the Pacific; for some it meant expansion over all North America, including Canada and Mexico; for others it meant the entire hemisphere, including Central America, South America, and the Caribbean. Later, the doctrine of Manifest Destiny was used to justify the acquisition of territories overseas.

Manifest Destiny, however, meant much more than mere expansion of American territory; it had political and socioeconomic elements that were essential to its appeal. Proponents of Manifest Destiny envisioned the creation of a society unlike any that had ever existed: a continental society that was middle class, avoiding extreme class divisions; a society in which freedom of religion prevailed; a society that was economically democratic, lacking monopolies and offering free land; a society that was a refuge for the oppressed, giving a new beginning to all who desired one; and a society having a democracy with wide suffrage and frequent elections. "A free, confederated, self-governed republic on a continental scale—this was Manifest Destiny," wrote Harvard historian Frederick Merk.

Few ideals in American history have had as much popular support as Manifest Destiny. Driven by the desire for cheap land and a fresh start, Americans moved west in mass. Throughout the nineteenth century, hundreds of thousands of men, women, and children headed for the frontier and became pioneers. When the newly independent Mexico opened its Texas territory to immigration, it expected a mixture of nationalities from around the world. Within five years, twenty thousand Americans had settled in Texas, forming the overwhelming majority of the population. By the 1840's, several thousands had settled the beautiful but remote Willamette Valley in Oregon; there would be fifty thousand by 1860. Although no more than eight hundred Americans lived in

California in 1846, fifty thousand to sixty thousand were moving there annually by the 1850's. And these were not the only boom states. By the early 1860's, Colorado's population reached one hundred thousand and Nevada's thirty thousand. After the Civil War, millions of farmers and their families poured into the Great Plains. As late as 1889, when Oklahoma opened its lands for settlement, twenty thousand massed at the state border; and by year's end it had a population of sixty thousand. Historians can multiply these few examples a hundredfold. Americans expressed their support for Manifest Destiny in the most convincing way: by their willingness to risk themselves, their families, and their fortunes to settle and develop new lands.

The nation's boundaries marched forward, because the dream of a free country stretching from the Atlantic to the Pacific was a centrist, moderate position supported by both citizenry and leadership. From 1791, when Vermont was admitted to the Union, until 1912, when Arizona was admitted, our republic expanded steadily with little opposition. State by state, the nation grew from a small coastal country of thirteen states to a continental empire of forty-eight. During these 120 years, the movement to expand was often controversial but never seriously in doubt. The opposition to the Louisiana Purchase in 1803, to the acquisition of Florida from Spain in 1819, to the admission of Texas in 1845, to the addition of the Oregon territory in 1846, and to the Mexican cession in 1848 was inconsequential. Manifest Destiny—meaning the settlement and development of the continent—was one of the most enduring and broadly supported programs in American history.

Every political idea has its radical forms. Manifest Destiny on a continental scale was a moderate position widely supported over a long period of time. International Manifest Destiny was not; it was an extreme version of traditional American expansionism and indistinguishable from classic imperialism. "I took Panama," boasted Teddy Roosevelt. The United States acquired overseas responsibilities after the Spanish-American War in 1898 and the construction of the Panama Canal in 1903. Both the Canal Zone and the territories Spain ceded—Guam, Puerto Rico, and the Philippines—were alien in language, culture, and political tradition to the United States. Continental Manifest Destiny raised neighboring peoples to equal statehood; American imperialism reduced distant peoples to colonialism. Other than a few brief episodes, Americans have never supported expansion beyond our present borders. Vietnam taught our leaders this lesson once more.

The time has come to declare a new vision of what America should become, because the dreams of the past have been fulfilled. "Our manifest destiny [is] to overspread and to possess the whole of the continent," wrote the charming and influential John L. O'Sullivan in 1845. But we have met that destiny. Our country extends from the Atlantic to the Pacific and forms a society unlike any the world has ever seen: a stable, self-governed, middle-class democracy on a continental scale. No wonder we stand out in late twentieth-century assemblies of constitutional states. Our ambitious ancestors intended to create an uncommon nation, and they have succeeded. But that mission is past. We now need a new ideology to guide us.

Space is our country's obvious destination. We are the principal heir of the most expansive civilization in history. We are the products of four centuries of frontier experience, experience that defines what we are and what we can do. Not for us are the imperial pleasures of conquest, subjugation, and rule; not for us are the triumphs of a sophisticated diplomacy pursued long and unremittingly. Our future lies on new worlds. Exploring, settling, and developing the solar system will call forth the best that is within us and place us in the vanguard of history: In space, we will culminate eons of expansion by life, half a millennium of expansion by the West, and centuries of expansion by our own nation. Stranded on Earth with a bumbling foreign policy and an enervated democracy, we will stagnate. We need a frontier to continue to be what we are—not an inept imperial Earth power erecting barrier after barrier to trade and immigrants to protect a declining prosperity, but a spacefaring nation with lifelines extending outward through the solar system, still a haven for the oppressed and still inalterably democratic.

The United States is not an ordinary player in the great drama in which humanity is involved. Neither the Russians nor the Chinese nor the Europeans nor the Japanese have that peculiar American combination of abilities our world needs to elude its troubles. America knows best how to create the self-sufficient communities that building an extraterrestrial civilization demands. We are the nation that knows how to liberate the energies of masses of people to work for common goals. We are the nation that knows how to rule ourselves without destroying initiative. We are the nation that knows how to invent and invent and invent until something finally works. We are the nation that knows how to avoid the chronic mistakes of colonialism. Meanwhile, time is running out. By no means is the rescue of our species assured. But these skills give our race a chance to build a solar-systemwide civilization and survive the dangers of overpopulation, environmental disaster, and nuclear war facing us.

Just as the principal occupation of Western immigrants in the nineteenth century was farming, so will the task of space settlers be to work the land. Unquenchable yearnings for their natural world will drive them to recreate the beauties of the mother planet. They will long for the green hills, blue skies, open fields, verdant forests, and running brooks that they left behind. They will demand prosperous settlements whose human landscape will more resemble small New England towns than New York and Mexico City. Space settlers will be importers of life from Earth. And in so doing they will build communities that will one day be as envied as America is today.

One day cities will cover the solar system in the same way they now blanket North America. These settlements will not be the urban disasters that pass for human habitations in our era. Twentieth-century megacities blister our fair planet because we have stayed here too long. Bottled up on Earth, we have no outlet for our energies. Gigantic urbs that agglomerate people and denature the environment multiply, part of the pathology of an Earth-based mankind. Such places are not possible off-world. To attract settlers willing to undergo hardship and danger, space cities must be true human communities that can provide full lives for all of their citizens.

Politically divided and overburdened with poverty, late twentieth-century Earth resembles fifteenth-century Europe before its explosion outward to the rest of the world. Humanity is ready to explode outward again, and the United States can lead that movement. The global crises, the need of our democracy for a frontier, our weakness as an international power, our position as heir of a great expansionary civilization, and the world's need for our extraordinary skills propel us toward space as our destiny. Failure to rise to this goal endangers ourselves, our species, and our planet. America as we know it has no future on a frontierless world, and neither, perhaps, does our civilization. If we embrace anew our heritage and choose to be what we are and to do what we can, we will again have hope. We will stand on the verge of a new phase of our history, one that may prove to be the greatest and most wonderful.

CHAPTER THREE

Preserving Our Nation: Domestic and Foreign Dangers

We live in an age that cherishes democracy. Only a rare government would admit that it was not democratic or striving to become so. Russia, as the Union of Soviet Socialist *Republics*, claims to be a representative democracy; and both mainland China and the autocracy in Libya call themselves people's republics. Americans are such lovers of democracy that they have invented new methods to measure the popular will. Opinion polls aid politicians in their votes on public policy; and consumer surveys guide businessmen in their marketing decisions. To the U.S. government, a test of a nation's loyalty to the West has often been its willingness to hold free elections. The support that the ideal of democracy enjoys in the twentieth century may exceed that of any other social value.

Because its rise to preeminence has been rapid and because its appeal is universal, we may not notice the obscurity from which democracy has risen. Before World War I, democratic governments ruled just North America and Australia; even Great Britain was an oligarchy in which less than 20 percent of the adult population could vote. Before this century, democracy never appeared outside Western civilization, and in most of history, it was only an episode of the ancient world, known mainly by scholars and antiquarians. "It is not easy to bring home to the men of the present day, how low the credit of republics had sunk before the establishment of the United States," wrote Sir Henry Maine, the famous British jurist and legal historian. After flourishing in ancient Greece nearly twenty-five hundred years ago, democracy was reborn after the French and American revolutions in the late eighteenth century. Serious observers believed that both were radical experiments, too unstable to endure. The survival of democracy in the nineteenth century was always in doubt. The French democracy lasted only a few years before Napoleon; and the outcome of the American experiment was not clear until after the Civil War. England may have been the birthplace of democracy in the modern world, but by the 1830's less than 5 percent of its adult population could vote. By 1880, that percentage had still not reached 10 percent; and the recognition of the right of all to vote did not come until the twentieth century. In the rest of Europe, the predominant form of government was monarchy until World War I.

The role of the United States in the development of democracy has been decisive. Although it is young as a nation, America has the oldest democracy on Earth. The government founded in 1776 was the first major attempt to rule a nation by constitutional democracy, and the right to vote was extended to all men decades before any other country. As the success of its political system became more and more apparent, America influenced the world profoundly. Millions immigrated to the land where the people ruled, strengthening global tendencies toward democracy. After fighting—in the words of the famous slogan—"to make the world safe for democracy," the United States emerged from World War I as a world power. President Woodrow Wilson won international acclaim by championing the right of national self-determination and urging the formation of an international legislature to adjudicate disputes. The League of Nations failed but was reincarnated as the United Nations twenty years later, when the great empires of the nineteenth century were falling apart because of global clamoring for self-rule.

If democracy fails someday in the United States, it will be a catastrophe for free men and women everywhere. The failure of democratic America would signal the end of an era. It would imply that free peoples cannot rule themselves and long survive among the family of nations. Our democracy should endure as long as so many other great political institutions—dynastic China, imperial Rome, and aristocratic Great Britain. The fall of democracy in the United States would encourage those who distrust free peoples. New royalists and new aristocrats would arise, despising all but members of their own class. The civil liberties we enjoy would fade, as the world entered a period in which a democratic government would be rare and unusual.

Unfortunately, we cannot be complacent about the security of America's democratic institutions. Students of history know that what is triumphant in one century may be reviled in the next. The hatred and contempt for colonialism in the twentieth century is matched by the high pride of the imperialists of the nineteenth. Contrary to the linear progression of science, political traditions are organic, cyclical and, by nature, impermanent.

Those of us born between 1946 and 1964 have inherited a democracy that faces unprecedented challenges. Domestically, the effects of the closing of the frontier in the late nineteenth century are finally being felt. America is entering a new era of limits that may lead to an increasingly stratified society and to weakening democratic traditions. Each of these effects will make our political life more divisive and less moderate, perhaps stressing our constitutional structure to breakdown. Internationally, the dangers are greater. The United States resembles an adolescent with introverted and irregular habits who has suddenly become king of a tumultuous domain. America has inherent weaknesses as an actor in world affairs, weaknesses that are the result of the very qualities we most prize. In a world that needs farseeing leadership, we practice global politics with no more than short-term purposes. Meanwhile, the crises worsen and worsen and worsen.

The solution to this problem is to break out into space. In a closed world in which the human race is penned with no hope of escape, the prospects for America are bleak indeed. Outside: the pressures of population, the maladies of the environment, and the dangers of war. Inside: a people with little experience as citizens in a democracy, other than mere voting; and a people plagued by increasing class stratification. As the decades pass, this country may begin to resemble a decaying fortress surrounded by impoverished mobs aching to have some taste of the riches within. If space does not become the focus of our national life, then our liberties, our wealth, and our security may be in danger. We need a new frontier to invigorate America's democratic traditions, to provide a safety valve for excess labor, and to be a breeding ground for new wealth and institutions. In international affairs, space will give us an arena in which our skills exceed other nations' as much as they fall short of them on Earth. Space will give the United States a new source of power to balance the power we constantly squander on this planet. We must lead the world in the exploration and settlement of space. We have no choice.

THE FRONTIER AND DEMOCRACY

In 1842, a man named Lansford W. Hastings was traveling west with 160 people from Independence, Missouri. Later, he recalled the intense feelings of harmony he felt toward the community with which he was journeying. He also recalled how his companions changed after they had traveled only a few days from the frontier. "All appeared determined to govern, but not to be governed. Here we were, without law, without order, without restraint. . . . Some were sad while others were merry; and while the brave doubted, the timid trembled! Amid this confusion, it was suggested by our captain, that we 'call a halt,' and pitch our tents, for the purpose of enacting a code of laws, for the future government of the company. The suggestion was promptly complied with.''

America's democracy and the advance of its frontier over a developing continent were inextricably intertwined.* The frontier alone did not make us a democratic people; other nations have faced empty lands during much of their history and not enjoyed democratic regimes. Our democracy also derives from ideas and institutions imported from Europe, such as the Magna Charta and the unwritten traditions of the English constitution. However, without the great western frontier, America could not have been America. Frontier life produced an egalitarian society and formed the free and independent character of generations of Americans. It created heroes who were neither aristocratic nor literary—the explorer, prospector, trapper, hunter, Indian-fighter, cowboy, covered-wagon pioneer, and farmer. In addition, the frontier moderated politics. Class-based parties failed to take root, in contrast to what happened in Europe. Low unemployment and rising wages kept labor peaceful, while better opportunities drew its leaders westward. Also, the opportunities for wealth in the West kept American life open and free from revolutionary social pressures caused by frustrated ambition. Talented men and women crossed class lines constantly and joined the established order. New institutions arose, preventing older ones from entrenching themselves and amassing ever more privileges. Finally, the bounty of the new lands fulfilled the dreams of the immigrant millions and won their allegiance to their adopted country. Without a frontier, America's democracy would

*As the first historian to write extensively about the relationship between America's frontier and her democratic institutions, Frederick Jackson Turner (1861–1932) was one of the most influential American historians of the last 150 years. His views, however, are controversial today. Historians do not accept without elaborate qualification most of Turner's "frontier thesis." But, as the Pulitzer Prize-winning historian Richard Hofstadter has written, "Even Turner's sharpest critics have rarely failed to concede the core of merit to his thesis, and wisely so." It is this core that is the basis for the arguments here, especially as it has been refined and developed by modern historians writing in the 1950's and 1960's.

have been little different from other democracies, and might have had an equally short life.

We are the products of a frontier society that stretches back to the earliest English settlements at the start of the seventeenth century, now almost four hundred years ago. A great void greeted the first settlers. Behind them was Europe, with many cities and a population of about 100 million. Before them was a continent twice Europe's size with a sporadic population of 3 million Indians and no large settled communities. In one of the great migrations of mankind, the American people spread over the continent and settled a wilderness. Generation after generation faced the frontier and lived by its harsh rules. In the beginning, there were vast forests, impassable mountains, desolate prairies, and barren plains. In the 1600's, the frontier lay in New England; by the mid-1700's, it had reached the Appalachian Mountains; during the 1830's and 1840's, pioneers poured into Michigan, Arkansas, Wisconsin, and Iowa; and by the 1880's, the West Coast had been settled and only portions of the Great Plains could be called a frontier. A raw wilderness had been converted into a world power. No one force, wrote the historian Ray Allen Billington, did more to Americanize the nation's people and institutions than "the repeated rebirth of civilization along the western edge of settlement during the three centuries required to occupy the continent."

Beyond the frontier lay experiences far different from those that formed the lives of those who did not go west. Living conditions were primitive; the population was sparse; and dangers common. Life was brutally hard. Backbreaking labor, disease, malnourishment, antisocial settlers, and hostile natives were unavoidable. At best, families lived a subsistence existence. On the other hand, the West was a place to earn a fortune and a reputation. Land, timber, oil, and minerals were cheap; and there were few obstacles to wealth. Men built mines, mills, and factories. New towns and cities appeared

with openings in business, the professions, and politics for anyone
knowing how to seize an opportunity. In the West, a new social order
developed, and it was open to anyone with energy and resource-
fulness.

The frontier became an agent of change. It was more than
new business opportunities, more than a savage place with a sparse
population, more than an everyday round of toil and deprivation.
It became a kind of magic box in which the older ideas and institutions of the East and Europe were subjected to the transforming
social conditions of a new environment. Civilization was always
beginning again at the edge of the wilderness. On a continually
advancing frontier, settlers were always forming new communities, returning to primitive economic and political conditions, and
beginning the slow climb to the complexity of urban life. This
process enabled the people of the frontier to escape from the class
hierarchy and the vested interests of the past. The frontier broke
old customs, creating new opportunities, new activities, and new
lines of growth; and it replaced old ideas and institutions with new
ones. Frontier society was a society in a state of perennial rebirth.

The frontier experience of building new communities molded
the character of generations of Americans into democratic citizens.
The absence of a prior leadership structure, wide dispersal of land
ownership, and the need for self-rule to solve unique local problems brought on a democratic spirit. Settlers learned how to govern
themselves. A new settlement lacks customs, clearly assigned
functions, and the accumulated physical equipment of life. Because it is composed of people similar in wealth and status, the
community has no existing leadership structure. Citizens must
work and survive without hospitals, churches, schools, and courts.
They face daunting problems such as brigands, Indians, plagues,
locusts, droughts, and fires. They must make new decisions, assume new roles, and serve as leaders. They may be called upon to
act as policemen, judges, priests, politicians, or teachers. Later

they must develop town life with market agriculture, credit institutions, retail trade, and transportation. All the avenues of enterprise such as manufacture and land speculation are open. Everyone experiences expanded roles, whether secular, religious, civil, or military. The citizens of these communities live a democratic life in the broadest sense. They share an activist attitude toward government, participate in public affairs much more widely than merely by voting, and exhibit a strong sense of personal competence in the management of affairs.

As a consequence, the successive Wests were more democratic than the settled East. For example, no state had complete manhood suffrage until the first frontier state—Vermont—was admitted to the Union in the late eighteenth century. And as more territories became states, they frequently vested unusual power in their legislatures, because they were considered more responsive to public opinion. The new western states also tended to rotate officeholders more rapidly by shortening terms of service and to extend the elective process to a wider range of offices, such as judgeships. Voters themselves displayed strong tendencies to elect citizens who had risen from the lower ranks of society. In addition, the devices of initiative, referendum, and recall that allowed popular control of legislation originated in the West. The frontier deepened the nation's democratic inclinations in many ways.

Successive generations of Americans lived through the same transforming discipline of frontier life. We conquered a series of wilderness environments and became accustomed to expansion and renewal. We developed a continent and experienced the same social evolution in a series of different frontier zones. Frontier life formed our character, conditioned our economic life, and nurtured our political institutions. In the harsh conditions of wilderness life, men were valued by what they could do—not by their class, education, wealth, or social connections. The frontier promoted economic independence and political equality. If a man disliked his circumstances, he could put his family in a wagon, head west, clear the land, farm, and win economic freedom. Free land created economic competency, which in turn created political power and an egalitarian society.

The frontier moderated political life in three ways. First, it prevented the development of an American proletariat and a serious left-of-center party. Instead, the settlement of the West led to a vast expansion of the middle class. Second, it provided a means by which the nation could assimilate the millions of immigrants pouring into the country. Their success and that of their children created a powerful constituency supporting American institutions. Finally, the sequential economic growth of the West guaranteed social mobility and continually created new institutions. American society remained open to energies and talents that might otherwise have been led by frustration to radical political action.

Ideological warfare, revolutionary change, and violence mar the political life of European democracies in the late twentieth century. Over forty governments have ruled the Italians since World War II, averaging only nine months in office. In the 1970's, the Red Brigades terrorized the country with political kidnappings, shootings, and kneecappings of prominent businessmen, intellectuals, and members of the judiciary. Germany has had similar problems with activist radicals. In France, a minor incident in Paris between the police and students in 1968 caused a wave of strikes that involved millions of workers and paralyzed the nation. Many feared that the French republic would collapse—France's fifth constitution during the period in which the United States has had just one. Only dramatic action by Charles de Gaulle preserved the government, and only after he had secured the support of the military. Great Britain is not an exception to the European tendency to violent ideological clashes. In 1981, Marxists gained control of one of its two major parties. The new leaders drove the moderates out, looked at Lenin and Trotsky as models, and sought a workers' democracy. They also advocated withdrawing from the Common Market, unilateral disarmament, and an end to the monarchy.

A striking difference between the structure of European and American politics is the presence in Europe of major parties that are based on ideology and class. The Communist party (ICP) is one of Italy's largest political parties. Although it is a proponent of a form of communism stressing independence from Moscow, it has never been part of a national government. Too many believe that the ICP does not accept democratic processes. Between 1951 and 1968, the French Communist party (FCP) won an average of over 22 percent of the vote and always had a large representation in the National Assembly. Yet this working-class party often acts as if it were an agent of the Soviet government. It is highly disciplined, highly centralized, and always endorses Soviet policy. The FCP was the only European party to support the Russian invasion of Afghanistan in 1979. The British Labour party originated in a congress of trade unions and has always had strong institutional and financial links to them. The Marxist left constitutes a large portion of the Labour party. It is far to the left of any socialist party that has governed France or Germany.

A strong left-of-center party never developed in the United States, because the settlement of the frontiers in the West deprived the working classes of any sense of solidarity or class consciousness. The frontier kept per-capita income high, siphoned off excess workers, and stripped labor of its leaders. As the West fed new slabs of rich resources into the developed sectors of the economy, per-capita income rose steadily. American real wages in the nineteenth century were substantially higher than those of the Old World. Moreover, the frontiers acted as a safety valve for the industrialized East. The laborer always had an opportunity to better himself by changing place and occupation. Consequently, labor was not radicalized as it has been in other industrialized countries. The West drew off union members and leaders in the same way that new industries draw off labor and entrepreneurs today. The frontier kept the percentage of the union labor force small and prevented the portion that was radical from exerting political leverage of any significance. Instead of becoming proletarianized, American labor became part of the middle class.

The thirty million immigrants who came to the United States between 1860 and 1930 strengthened America's democracy. The foreign-born and their children dreaded political change because they feared altering the social order and disrupting the family. They stubbornly opposed the proselytizing of socialists and radicals. Each reformers such as Theodore Roosevelt, William Jennings Bryan, and Robert La Follette failed to attract immigrants. The new arrivals supported their adopted political system. They may have done the grueling work that developed the resources of the nation rapidly and cheaply, but they also found the personal freedom and economic opportunity they had sought. Immigrant letters home were overwhelmingly favorable, and remittances to relatives fueled the continuing stream of immigration.

Stratified societies that limit movement between social classes can create reservoirs of able yet discontented men. For example, because non-Russians never rise very high in party or governmental hierarchies in the USSR, experts believe that ethnic minorities there have the potential for disrupting Soviet life. The colonial aristocracy before the American Revolution, the bourgeoisie before the French Revolution, and the proletariat before the Russian Revolution were shut out from positions of political power. Privileged groups ruled these countries and excluded the talented of other classes. When the rebellion began, many natural leaders came to power with no sympathy for the existing regime.

The frontier guaranteed the United States the open-class society implied by the Constitution. Americans have always been able to move up and down the social scale, creating new wealth and institutions. Each era has witnessed the birth of new fortunes and new men. In the eighteenth century, it was merchants such as Thomas Hancock and John Jacob Astor; in the nineteenth century, it was industrialists such as John D. Rockefeller and Andrew Carnegie; in the twentieth century, it was manufacturers such as Henry Ford and Walter Chrysler; more recently it was technologists such as William Hewlett and Edwin Land. Nor was success confined to a few dramatic examples. In his famous essay *Self-Reliance*, Ralph Waldo Emerson wrote of the typical Yankee going west. Starting as a farmer, he became a storekeeper and then a land dealer. Later he practiced law, served in Congress, and became a judge. Emerson's characterization was not far from the truth. The frontier put few restrictions on what a person could do.

In 1890, the Census Bureau announced that the frontier as a continuous line had ceased to exist. Nearly twenty-five years later, the historian Frederick Jackson Turner addressed his generation:

> For three centuries the fundamental process in [our] history was the westward movement, the discovery and occupation of the vast free spaces of the continent. We are the first generation of Americans who can look back upon that era as a historic movement now coming to its end. Other generations have been so much a part of it that they could hardly comprehend its significance. To them it seemed inevitable. The free land and the natural resources seemed practically inexhaustible. Nor were they aware of the fact that their most fundamental traits, their institutions, even their ideals were shaped by this interaction between the wilderness and themselves.

These words may also be directed to the generations alive today. Although the frontier as a continuous line had disappeared

in 1890, the open West endured until 1934, when Congress passed
the Taylor Grazing Act and ended the era of public-land distribu-
tion by homesteading. Moreover, the frontier as an open form of
society, in which new social structures arise, reappeared many
times in this century. Each time a new area of the country was
developed, Americans relived experiences that they have under-
gone since Jamestown was laid out in 1607 with a fort, a church,
a storehouse, and a row of little huts. Opportunities abounded for
talent and energy; new men and institutions were thrown up; new
wealth was created; political power was tasted by new groups; and
the old order had to adjust to accommodate the new. Hollywood in
the 1920's, the Texas oil fields in the 1930's, southern California
in the 1950's, and Silicon Valley in the 1970's are the most spectacular examples of this recurring phenomenon.

But in the late twentieth century, such experiences have become rare. Partly because of the huge size of the newest generation
and partly because the physical limits to the country have been
reached, most endeavors are already crowded with competitors.
The amount of capital, the number of credentials, and the time of
preparation needed to enter a field has escalated. In the early years
of the Republic, young Americans could enter agriculture or trade
with little formal training and become leaders of their communities. Such accomplishments were still possible in the late nineteenth century, when over half of the multimillionaires reaching
maturity in the 1890's originated in the lower classes. But by
1950, that percentage had dropped to less than 10 percent. Today,
franchises are gobbling up the small-business sector of the economy, while the new frontiers are advanced areas of technology
such as genetics or electrical engineering—fields requiring a
lengthy and specialized education. Even political careers have become less open to newcomers because of the increasing cost of
campaigns and the high reelection rate of incumbents. In 1986, the
reelection rate for members of the House of Representatives topped

98 percent; and at the end of 1987, House members had accumulated a total of $64.5 million for the upcoming election, while their challengers had raised only $1.5 million.

For us Americans living in the late twentieth century, the frontier has closed. Thinkers and politicians speak about a new era of limits; the young are warned that their standard of living will be less than that of their parents; and in a country where land was once free for the taking, stratospheric prices are pushing home ownership out of the reach of many. Whether this closing occurred in the nineteenth century or some time in this century is immaterial. The crucial facts are these: No longer do the experiences of democratic community life mold the characters of Americans. Gone, too, are the moderating effects that the frontier exercised on our political life. The conditions in which our historically fragile form of government developed have disappeared, perhaps forever.

Other than as voters, few Americans today participate as citizens in a democracy. Nameless bureaucrats and remote politicians make the decisions that govern the way we live. The most common political experience we have is that of an employee in a hierarchical corporation where democratic practices are infrequent. Ordinarily, we dwell in the privacy of our economic and social lives, rarely emerging into the expanses of public service. Meanwhile, election campaigns lengthen and become exorbitantly expensive, as media consultants and professional campaign managers usurp what was once the part-time pursuit of citizens. No wonder voter registration is low, cynicism concerning our leaders rampant, and alienation from politics normal.

Class lines are beginning to form. The income gap between the richest families and the poorest is now wider than it has been at any time since the Census Bureau began keeping such statistics in 1947. Although the top 40 percent of income earners gained slightly at the expense of the bottom 40 percent between 1947 and 1972, the real change started in the mid-1970's. Between 1976 and

1986, the rich became richer, the poor became more numerous, and those in the middle did less well than before. In 1968, the poorest fifth (20 percent) of families with children received 7.4 percent of the total income of all families; in 1983, their share was only 4.8 percent, down by a third. In 1988, the top fifth of Americans accounted for 43 percent of the national income—the largest share since World War II—while the bottom fifth accounted for 4.7 percent—the smallest share in the same period. Some economists have predicted that the middle class will disappear altogether, leaving the country torn between an affluent minority and the desperate poor. The polarization is already beginning in the retail industry. Mass merchandisers such as Korvettes and Gimbel's are gone, while Sears, Roebuck, & Company and J. C. Penney are repositioning themselves for an upscale market. Bloomingdale's and Neiman-Marcus prosper by catering to the affluent; K mart and Woolco do so by selling to those constrained by poverty or thrift.

Evidence of a divided America has begun to surface. According to the eminent Harvard sociologist Daniel Bell, the idea of the middle class has begun to break up. More and more young people cannot buy homes. A typical thirty-year-old in the 1950's had to spend 14 percent of his paycheck to make payments on a home. By the 1970's, the required share of income rose to 21 percent, and by 1985, to 44 percent. In 1950, a family with a father at work and a mother and two children at home represented 70 percent of the labor force. Today, that figure is 15 percent, because of the rising cost of home ownership and the movement of labor into the lower-paid service sector of the economy. To compete for scarcer resources, interest groups form and polarize the nation by region, education, and age, while middle-class values fade.

The passing of the frontier means that the number of economic niches for workers and entrepreneurs is dwindling. Since the United States no longer has a safety valve for its excess labor, the level of unemployment is rising. In the 1980's, the Reagan administration was proud that it had reduced unemployment to less than 6 percent, although millions of Americans classified as employed are underemployed, or are working at jobs that are vulnerable in a recession. Perhaps the English economy foreshadows our own. British unemployment in late 1986 was over 17 percent, in terms comparable to those used in the United States. Small business has always been a refuge for those seeking autonomy. Unfortunately, the franchise industry has been growing at an average rate of 10 percent a year, and government forecasters believe that by the year A.D. 2000 two thirds of all retailers will be franchises. Franchises are quasi-corporate units. Franchisors rule their franchisees with strict policies and slice off a hefty percentage of the profits.

Immigration to the United States continues in large numbers, especially across our long southern border; and the new arrivals are straining the capacity of this country to assimilate them. For example, southern Florida has become Latinized. Angry natives resent not being able to find a job or buy groceries in Miami unless they can speak Spanish. Black leaders fear new outbreaks of violence in slums by blacks shoved aside by waves of Cuban immigrants. Experts believe that the growth of illegal aliens has accelerated in recent years; there may now be as many as 12 million illegals living in the United States. Population pressures and political turmoil will drive many more South Americans northward in the coming years. According to Census projections, over 25 million Hispanics will live in the United States by 2000, a jump of nearly 46 percent.

A symbol of the times is the transportation problem that has emerged in the West. California was once open land where travel between two points was free and direct. No longer. On Los Angeles freeways during commuting hours, traffic often slows to 35 mph and even to 15 mph. Engineers, professors, and urban planners agree that the situation is not likely to improve. Congestion is intensifying faster than the population is growing. A rush-hour ride on the San Diego Freeway that now takes one and a half hours may take three hours by the year 2000. Even on mornings when no accidents occur, about three hundred miles of freeway in Los Angeles are lost to congestion. In 1963, only thirty miles were backed up on a typical day. The whole country is becoming similar to California freeways. Everything is structured; everything is developed; everything is occupied; free movement is impossible.

A French politician, Alexis de Tocqueville, wrote one of the most perspicacious and profound analyses of American institutions, *Democracy in America*. De Tocqueville's masterpiece remains the classic interpretation of democratic and egalitarian institutions in this country. He foresaw with remarkable accuracy the influence of democratic life on religion, the family, business, the arts, science, and the military. He also believed that "a boundless continent" that was "open to the exertions" of Americans was "the chief circumstance which has favored the establishment and the maintenance of a democratic republic in the United States." Now that we no longer have an open and boundless continent, will we lose our democracy because we do not know how to act as free citizens? Will our politics acquire a European violence and instability, as class lines solidify and unassimilated immigration continues? Who can say? All we know is that an enduring democratic government in the great drama of human history may be unique, and that some of the conditions that favored the American democracy have passed. The frontier nurtured our democracy, and now it is gone. "This then is the heritage of pioneer experience—" wrote Frederick Jackson Turner, "a passionate belief that a democracy was possible which should leave the individual a part to play in free society." How can a nation that has no place for pioneers keep that passion and that belief?

THE IMPERIAL PROBLEM

The British Empire began with the first Virginia Colony in 1607 and ended with the independence of India after World War II. It included Great Britain herself, Northern Ireland, Canada, Australia, New Zealand, India, Pakistan, and a half-dozen African countries. At its peak in 1918, the British Commonwealth of Nations embraced a quarter of the earth's surface and a quarter of its population. It made many contributions to civilization. The renowned imperialist Winston Churchill argued that it "has spread and is spreading democracy more widely than any other system of government since the beginning of time." Some historians believe that British institutions such as the military academy Sandhurst and the London School of Economics destroyed the empire, because they taught generations of Africans and Asians to think for themselves. British imperialism was also an agent of change. It injected new ideas into torpid societies, shook up stagnant cultures, and everywhere laid the foundations for industry.

The United States also heads an empire. It includes Europe, almost all the Western Hemisphere, and important parts of Asia (Japan, Australia, New Zealand, South Korea, the Philippines). The American Empire is an economic community tied together by trade and protected by military force. The United States holds this preeminent position for the same reasons all imperial nations have ruled: wealth and power. Since the end of World War II, America's military and economic might have overawed its allies. We may not call our realm an empire, we may not have an emperor, and we may have less control over subordinate states than past empires—but the Western Alliance is an empire nonetheless. It is protected by American missiles and enriched by the American market. For all Western heads of state, the path of honor and influence leads to Washington, D.C., and the White House. Who "can doubt that there is an American empire?" writes the influential historian Arthur Schlesinger, Jr.

The sway of the United States in the world extends beyond its allies. We have been Earth's greatest power since 1945, when we had a nuclear monopoly and an industrial plant undamaged by World War II. Over forty years later, the United States remains the dominant power on the planet. The structure of international politics since the end of World War II demonstrates our profound influence on world affairs. America dislikes colonial empires, supports national self-determination, and champions democracy. Consequently, after World War II, the overseas empires of the Europeans were disbanded and the number of independent nations multiplied. Democracy—if only in name—became the favored form of government, no matter how tyrannical. The United Nations itself is a symbol of American domination of the era. The UN is a child of the United States. It has a quasi-democratic structure, with legislative and executive branches; it is headquartered on American soil; and throughout most of its life, it has received about half its financial support from the United States. It is hard to imagine an institution such as the United Nations existing in any era not dominated by a democracy.

The imperial problem of the United States arises from the conflict between our democratic ideals and our role in world politics. America suffers from the predicament of being both a democratic nation and an imperial ruler. Within our borders, we have a political life in which democratic practices prevail; outside, our position in the world often requires us to make decisions and to take actions that violate our democratic traditions. Too often we cannot be true to our heritage and at the same time act in accord with national interests. This conflict endangers our position in the world. How we can preserve our democracy and survive in the competition among nations is the imperial problem of the United States.

Our form of government is not the only reason the United States is weak in international politics. We also suffer limitations because our experiences with other nations have been few and atypical. It is astonishing that the first years of the United States as Earth's greatest power after World War II were also the first years of our continuous involvement with the rest of the world. Before the twentieth century, the Pacific and Atlantic oceans insulated us from the tumults of Europe. We entered World War I a year before Germany surrendered, and returned to isolation as soon as the war ended. No wonder our actions often appear awkward and bizarre to our more worldly allies. Our people and our institutions have not had centuries to become accustomed to the intricacies of international affairs, as theirs have had.

Two dangers threaten us. The first is the rise of a traditional world power with the economic strength of the United States without the distracting effects of a democratic political life. If China modernizes its economy, it would become a far more formidable rival than is Soviet Russia today. The second danger concerns the global crises of overpopulation, ecological disaster, and nuclear war. These crises may transform the world is such a way that international politics as currently practiced by the United States becomes impossible. If a Third World nation detonates an atomic bomb in a Western city, American reluctance to intervene directly in the affairs of other nations may disappear—and with it much that is precious about this country.

Long periods of peace often characterize eras governed by successful ruling powers. The *Pax Romana* was a time of about two hundred years in which the world dominated by the Roman Empire enjoyed peace. The Romans ruled an area stretching from Iran to Scotland. Similarly, the Ming dynasty (A.D. 1368–1644) was one of the most stable and most powerful dynasties of Chinese history. It collected tribute from countries as far west as the African coast and as far east as Japan. For more than a hundred years, from the 1420's to the 1530's, its empire was tranquil and prosperous. Historians sometimes call the nineteenth century *Pax Britannica,* because the British Empire established the rule of law in many turbulent places and kept peace for the better part of a century.

The possibility of a *Pax Americana* is remote. The United States is an ambivalent ruling power, in love with democracy and persistently isolationist. Our brief and unrepresentative experiences have not prepared us and our institutions for world leadership. We have been a democracy far longer than we have been an empire; we are more interested in human rights and free governments than in a stable international order. Unfortunately, principles such as peace and national self-determination may be incompatible. The global record of conflicts since World War II, when U.S. hegemony began and multiplied the number of independent states, is revealing. Since 1945, 50 million have died in nearly two hundred wars, creating another 50 million refugees. The United States has buried over 110,000 soldiers and hospitalized a quarter-million more in the same period.

The idea that democracies are unskilled in the conduct of foreign affairs is not new. The French politician de Tocqueville itemized their disadvantages over 150 years ago. He wrote that a democracy cannot "persevere in design" or "combine its measures with secrecy" or "regulate the details of an important undertaking" or "work out its execution in the presence of serious obstacles" or "await their consequences with patience." We do not have to search widely to find examples of these characteristics in American history. The United States is a rough-and-tumble democracy that permits countless factions to influence policy. We believe in a limited government and rotation of elected officials. We prefer, in the words of Woodrow Wilson, "open covenants openly arrived at" to secret diplomacy. These practices form part of a precious political heritage that preserves our liberties and our free institutions. The weaknesses de Tocqueville noted result from political traditions that we cherish. We cannot eliminate them without ceasing to be a democracy.

The involvement of the United States in world politics has been brief, and its experiences have been peculiar. Most countries experience the perils of international affairs early in their national lives, because they must fight for survival with nearby rivals. England, for example, warred for centuries with Wales and Scotland before uniting with them in the larger nation of Great Britain. Beginning as a single city, ancient Rome first conquered its immediate neighbors and then the entire Italian peninsula. By the time it was competing with regional powers in the Mediterranean, it had had centuries of international experience. The development of Europe, the USSR, and China has been similar. The history of the United States has been different. For a ruling world power, we have been active internationally for only a short time, and our experiences have been very unusual. We never learned how to act as an international power. American culture, as a result, does not synchronize with American responsibilities. What the United States is called to do and what its people will support do not match.

Long years of struggle, defeat, and triumph never taught us how how to deal with rival powers. Only after World War II did the principle that the United States had vital interests overseas become a part of our political consensus. Compared to other nations, compared to other world powers, and compared to our own history, America has been involved in international affairs for only a short time. In world diplomacy, the United States alone admits candidates into its foreign service without language requirements. Almost one quarter of recently promoted senior foreign-service officers lacked fluency in any foreign language. Our education in diplomacy has just begun.

Our experiences during this period have also been peculiar. Success in foreign affairs demands steady application to a series of frustrating and complex problems. Yet our initial experiences accustomed us to easy victories and long respites from responsibility. The Spanish-American War began in 1898 and was over in ten weeks. Few Americans died, and we won the entire Philippine archipelago and Puerto Rico. In World War I, we fought in France for only eighteen months and provided the margin that won the war. Europeans showered us with affection and honor. World War II was a much greater ordeal, but our experience was different from other nations'. World War II for us lasted less than four years and—other than Pearl Harbor—did not touch our cities. The English, Italians, French, Germans, and Russians were at war for over five and a half years. Their cities were bombed mercilessly, and civilians died by the millions, sometimes at the hands of their own countrymen. When the war ended, their lands were in ruins.

Because we spent our national youth in isolation and because our experiences as a new-fledged power distorted our view of the world, the United States does not play the role of a superpower well. Some have become convinced that we are unable to conduct a serious foreign policy because of the nature of our society and our political institutions. Our leaders cannot depend on us for support, and they often lack qualification for their responsibilities. A majority of Americans, according to the twenty-five-year-old, nonpartisan Atlantic Council (a Washington, D.C., think tank in foreign affairs), are "relatively illiterate in international affairs." The Atlantic Council's opinions are not disputed. More people *teach* English in the Soviet Union than *learn* Russian in the United States. The council reported in 1987 that two thirds of the voters take little or no interest in foreign policy, although 17 percent of production workers owe their jobs to international trade, 40 percent of U.S. farmland produces for export, and 33 percent of U.S. corporate profits come from international activity. Given this apathy, it is not surprising that the United States is unable to build a lasting consensus that can support a consistent foreign policy.

The handicaps of our leaders also weaken the United States. Presidents are generally inexperienced in international affairs. Typically, the background of presidential candidates and their followers lies in domestic policy and national politics. They must always relearn the lessons of their predecessors. Furthermore, turnover plagues the foreign-policy process. Between 1976 and 1986, there were four secretaries of state; and Ronald Reagan had five national-security advisers in six years. The civil service assists little. The State Department is the smallest U.S. Cabinet agency. Budget-cutting in the 1980's has swelled workloads, leaving the foreign service underfunded and understaffed in key areas.

Providence has blessed this country and its citizens with many wonderful gifts, but skill in international affairs is not one of them. Although we may be brilliant in the practical matters of business and technology, we are much less versatile in our dealings with other nations. A democratic foreign policy cannot be covert; it cannot be controlled by an elite of experts; and it cannot remain unaffected by the policies of newly elected officials. Nor can we escape from the effects of our history. We are not internationalists. American travelers are not famous for their desire to learn foreign languages, and American businessmen are not admired for their eagerness to export products. The pressures of world rule may be changing us, but we still prefer the pleasures of isolated quiet. The price we pay is a comparative weakness in world affairs, a price we can afford as long as we retain our dominant position in the world.

The country most likely to challenge our leadership is China. Not only does this remarkable country possess abundant natural resources and a huge population, but it also has a history of achievement in government that is more than two thousand years old. Moreover, it is part of a family of nations that have demonstrated a talent for adapting to the modern world.

The rise of East Asia to wealth and power is one of the most remarkable political and economic phenomena in the latter half of the twentieth century. About 200 million people in five Asian nations—Japan, South Korea, Taiwan, Hong Kong, and Singapore—made a successful transition from peasant agricultural societies to modern industrial states. The achievement of Japan, whose cities were nearly destroyed in World War II, is representative. Its gross national product grew by a factor of fourteen between 1950 and 1980, averaging 9–10 percent growth per year.

China may be on the verge of an economic and social transformation similar to those that have occurred elsewhere in East Asia. It shares many important characteristics with Japan, Taiwan, South Korea, Hong Kong, and Singapore. Each of these nations has a stable government. After the disastrous final years of Mao Tse-tung's reign, China appears to have established some of the conditions that were factors in the growth of its neighbors' economies. Each of these nations has a talented and industrious people. Experts believe that China has a human resource base in the 1980's as strong as that of Korea and Taiwan in the 1950's. Each of these nations has an economic policy that reduces income inequality and emphasizes experimentation, manufacturing, and exports. China's policies are the same. Its economic achievements by the early twenty-first century may make those of Japan today seem small. China has already demonstrated its economic strength by growing at over 8 percent per year from 1977 to 1985.

Two examples of the national resolve of the Chinese have been the success of their agricultural and birth-control policies. Many Third World countries remain poor and grow poorer because they cannot control their populations and cannot become agriculturally sufficient. Because of its huge population, the magnitude of these problems for China is much greater than for smaller nations.

Yet the Chinese have conquered them. They reprivatized agriculture, becoming the largest grain producer in the world and a net exporter of food. (China exports to the Soviets, who remain food-importers after decades of five-year plans.) China also curbed its population by introducing in 1979 a program of incentives for late marriage and penalties for having more than one child. An American demographer believes that some 250,000 infant girls were killed from 1979 to 1982 by parents who preferred a boy to a girl. Such is the power of the Chinese state.

We can find similar demonstrations of ability in government throughout the history of China. In fact, Chinese civilization raised the art of government to levels matched by few other nations in history. "Here they have been among the most successful of all the peoples of the globe," wrote the Yale historian Kenneth Scott Latourette. The French historian Jacques Gernet believed that one of China's "outstanding merits is to have developed, in the course of a long process of evolution, complex forms of political organization which were the most highly perfected in the history of human societies." Except for occasional periods of disorder—such as in this century—Chinese dynastic government extends back over twenty-two hundred years. The first Chinese emperor began his rule about two hundred years before Christ; the last emperor was deposed in the second decade of this century. Western civilization would equal this record if the Roman Empire ruled Europe today instead of having fallen about fifteen hundred years ago.

The West cannot discount this extraordinary achievement by pointing to a preexisting cultural and racial unity. China was always more difficult to unify than Europe—the birthplace of Western civilization—because it was larger in both area and population. Like Europeans, the peoples of China are racially mixed; they are the product of a constant mingling of races. Language divides the Chinese as much as it does the Europeans; the so-called dialects of

China differ from one another just as markedly as do the European languages. Chinese culture is also as diverse as the cultures of Europe. If we see in China in the twentieth century a nation more uniform than most, we are observing the success of centuries of government under the influence of Confucian philosophy. By adopting policies that did not discourage internal migration or intermarriage among classes, the Chinese government produced a nation with an unusual degree of cultural and racial unity.

For over three decades, Norman Cousins was editor of the *Saturday Review,* a widely respected weekly with a circulation of nearly half a million at its peak. He acted as an unofficial ambassador for President John F. Kennedy in negotiating the nuclear-test-ban treaty and has written numerous editorials, hundreds of essays, and more than a dozen books. Cousins believes that the world's social and political institutions are incapable of resolving the problems of nuclear war, environmental deterioration, and population growth. Item: None of the three principal tropical-rain-forest nations—Brazil, Indonesia, and Zaire—has a program to reverse the rapid devastation of the forests that imperils the quality of the earth's atmosphere. Item: Libya, Iraq, Iran, and North Korea have attempted to acquire technology to build their own nuclear weapons. It is not available in the 1980's, but it may be in the 1990's. Item: Explosive population growth in Central and South America may overwhelm the capacity of governments to adjust adequately and cause instability of major political significance to the United States. For Norman Cousins, the obvious solution lies in creating supranational institutions constituting a form of world federal government. "The big news of the 21st century," said Mr. Cousins, "will be that the world *as a whole* has to be managed, and not just the parts."

If the global crises that loom before us create such great disorder that a central world authority becomes a political necessity, a newly industrialized China may have a better claim to world rule than the United States. The basis of our political life is the Constitution, a document that was designed for a seaboard farming nation of mostly British background. It now accommodates the needs of a continent-spanning country of 240 million that has a citizenry of great ethnic and religious diversity, an economy built on advanced technology, and responsibilities that are global. Can the Constitution now be stretched to the extreme of world rule without harming its fundamental spirit? Can an instrument dedicated to fulfilling the rights of individuals to "Life, Liberty and the Pursuit of Happiness" serve as an ideology for a global empire? Can a people that champions democracy and dislikes intervention in the affairs of other countries head a new world order that must be dictatorial? Such concerns will not trouble a Japanized China, with more than 20 percent of the world's population and an imperial tradition two millenia old.

In some ways America resembles the ancient city of Athens. Until the United States was formed in the late eighteenth century, the fifth century B.C. Greek city had been the world's greatest democracy. We remember Athens in its glory as a golden age in which a free people created enduring monuments of the intellect and the imagination. The work of Plato, Aristotle, Aeschylus, Sophocles, Euripides, Thucydides, Herodotus, and others will continue to inspire and educate us as long as we value our humanity. The twentieth century has been a brilliant period of innovation for the United States. The technologies of communication, computers, and aerospace alone will affect future generations as deeply as Greek civilization has affected us. Let us hope that the similarity ends here, for Greek political skills led Greece into bondage to foreign powers. Just as the United States seeks to spread democracy to other nations, Athens sought to install democratic governments in other cities in Greece. But she failed to unify the country under her rule; subsequently, the monarchy of Macedonia and then the oligarchy of Rome conquered her. Greece lost her freedom for centuries.

In the latter part of the twentieth century, the United States is the richest and most powerful country on the earth. No nation matches the combination of our wealth, our power, and our influence. But this preeminence may be transient. One hundred years ago, the sun never set on British dominions; fifty years ago, the Japanese were conquering Asia while the Germans were mastering Europe. Twenty-five years from now, the world scene and its important players may change again. China has the population, the resources, and the culture to become a fearsome rival to the West in the twenty-first century. We may be witnessing the emergence of a new Chinese Empire, one armed not only with modern technologies but also with the political traditions of an ancient civilization.

Preserving America's peace and prosperity in the coming decades may be as great a challenge as any we have faced in our history. We must contend with formidable foes and with global crises that threaten to overwhelm us. Unfortunately, we confront these problems at a time when our people, our leaders, and our institutions are still adapting to a world in which the isolation of our national youth is no longer possible. But our greatest task will be to protect the political traditions we treasure. Fierce national rivalries, the demands of a world run amok, and the seductions of great power will often tempt us to compromise our heritage of democracy and human rights. Decisions will never be easy, because the choice will usually be between actions that serve real economic or military interests and those that are humane. We will lose our country if we fail to cultivate our national interests; and we will lose ourselves if we forget to honor in practice the principles on which America is based. How to survive and prosper while supporting democracy and furthering human rights is the imperial problem of the United States.

THE SECOND FRONTIER

The launching of the space shuttle reopened the American frontier. On April 12, 1981, the *Columbia* blasted off from Florida—one of America's first frontiers—and landed fifty-four and a half hours later in southern California—one of America's most recent frontiers. The world sensed the significance of what was happening. More than a quarter-million people traveled to the remote airstrip in the Mojave Desert to watch the *Columbia* return to Earth; hundreds of millions more watched on television. The successful maiden flight of the *Columbia* began a trail to the heavens that will one day become a highway.

Space is America's second frontier, as the West was its first. "He would be a rash prophet," wrote Federick Jackson Turner after the western frontier had closed, "who would assert that the expansive character of American life has now entirely passed." He would indeed! Our vitality as a people is linked to the presence of a frontier in American life. It is the characteristic of this country that differentiates it most from the short-lived democracies that preceded it. The West fulfilled the dreams of immigrant millions; it raised labor into the middle class; it kept America egalitarian; it gave opportunity to the talented and the industrious; and it transformed a small English seacoast colony into a multi-ethnic world power that contains much of the best that humanity has to offer. The opening of the space frontier continues the great traditions of our history. It offers the hope that the American Dream will continue to serve as a vision of hope for the unhappy and the oppressed of all lands. "The opening of a new high frontier," wrote Gerard O'Neill, "will challenge the best that is in us" and "give us new freedom to search for better governments, social systems and ways of life."

A destiny in space fits America. It will help preserve our democracy, it will balance our weaknesses in international affairs, and it will allow us to do what we do well. Stable, enduring democracies are unknown in history. And an open space frontier will help ensure the preservation of our political heritage by maintaining the conditions out of which these traditions arose. Space will also strengthen us internationally. We are not an imperial country. We have neither the heart for conquest and rule nor the aptitude for complex international rivalries. We are instead explorers of unknown territories, builders of new communities, and developers of frontiers. Our skills are far more suitable for settling the solar system than they are for conducting foreign policy on Earth, and we will be more successful planting democracies on other worlds than we have been spreading our political ideals on this planet. Locked up on Earth, frustrated by closed frontiers, our great talents will waste away and our democracy may die.

We will never be able to democratize a planet with cultures as diverse as an impoverished Africa or an Islamic Middle East or a Soviet Russia. Nor will human rights ever flourish in a world oppressed by unwanted hundreds of millions and by the sickening omnipresence of nuclear war. But space is different; space is new. The vast and wondrous realms outside this mere planet can refresh our species as a child's embrace rejoices a parent's heart. America's movement into space and expansion over the solar system can further the immortal causes of democracy and human rights in a way than can never be duplicated on Earth. America's future lies in space. It is this future that the gifted generation must open up.

CHAPTER FOUR

The Great Opportunity Before Us

Once, a very long time ago, the land on this planet was bare. There were no animals, birds, insects, trees, bushes, flowers, or grasses. Earth's landscape was rocky and bleak because life was confined to the oceans, where it had been flourishing for billions of years. Those ancient waters contained a vast multitude of creatures, including vertebrates, mankind's ancestors, and representatives of all other animal phyla. But when life emerged from the sea, its possibilities enlarged on a magnificent scale. The age of plants, the age of reptiles, the age of mammals, and mankind's shining moments lay in the future. The immensity of oceanic life was dwarfed by what was before it on land.

Today, the land has been filled, and life stands at a similar juncture. Before it lies a galaxy of stars and planets that are its future homes. Life, represented by humanity, is on the verge of another transformation. To the faint intelligences of those creatures who first emerged from the oceans, the land was also a foreign realm where survival was difficult and uncertain. As the other suns in our galaxy appear to us, so the huge expanse of land must have appeared to them: unknowable and without end. So also was their future—beyond imagination and beyond dreams—the same as ours.

The drive impelling man to the stars is an ancient one. As has been noted, life was purely an oceanic phenomenon for billions of years until it exploded onto land, rapidly covering the continents with a copious variety of plants and animals. The next advance was into the atmosphere. The existence in the modern world of nearly ten thousand species of birds attests to the success of that expansion. Mankind's own history has mirrored that of life itself, as we moved from our birthplace in eastern Africa to every habitat on the planet. The relative speed of our victory on Earth is remarkable. *Homo sapiens* is a very young and vigorous species. Our individual lives may be short, and the generations of our ancestors may stretch out endlessly behind us, but humanity with our present physical and mental endowments is scarcely one hundred thousand years old. Mankind has been living in cities and able to write for no more than ten thousand years. Many common animals are much, much older. Horses, birds, snakes, and sharks represent species millions of years old. The age of dinosaurs—an earlier, globally dominant animal—lasted over 200 million years. In a very short period of time, we have learned how to dominate this planet.

The next logical extension of life—after the seas, the land, and the atmosphere—is into space. From a biological point of view, the expansion of life into the solar system carries an inevitability and a momentum reaching back to life's origin. One thinker—the scholarly polymath Krafft Ehricke—labeled this tendency an "extraterrestrial imperative." Others have compared humanity's sojourn on this planet to a baby's stay in its mother's womb. In the same way that an overdue birth makes the mother ill, mankind is fouling this planet because his birth into the maturity of extraterrestrial life is overdue.

TEEMING PLANET, UNIQUE SPECIES, EMPTY GALAXY

If distant intelligences were examining the Milky Way, our galaxy, they would detect a stark contrast between what is happening on Earth and what is apparently *not* happening anywhere else. An unusual form of life threatens to overpopulate the planet. Our species, *Homo sapiens,* is growing so fast into every available habitat that we are causing an unprecedented number of extinctions. Humanity's conquest of Earth would perhaps be less surprising if we were an ancient and venerable species that had developed its powers over a long period of time. Yet mankind is very young: Our line is a mere babe compared to the lines that we are eliminating. We have achieved so much so quickly because of a talent for expansion unmatched by any other known form of life. And we have already begun to send out probes to explore other planets orbiting our star. Although life may exist elsewhere in the galaxy, there is no sign of any creatures similar to us. We appear to be alone.

World population has been soaring for more than a century. In the early nineteenth century A.D., it reached 1 billion. A little more than one hundred years later that figure had doubled to 2 billion, and the third billion was added only a few decades later. It reached 4 billion in 1974, and 5 billion in 1986. Conservative forecasts place population by the year 2050 at over 8 billion. Other experts expect it to reach 7 billion by 2000 and exceed 12 billion by 2050.

Extinction is a natural phenomenon that has occurred many times in the course of evolution. Wherever a new species becomes dominant in a region, it gradually eliminates the less fit species, while its own population explodes. We are now witnessing this process on a global scale. *Homo sapiens* is eliminating other species at an extraordinary rate. Although the number of extinctions averaged only seventeen per century between 1600 and 1980, biologists predict that the average between 1980 and 2000 will shoot up to 145. In 1984, the U.S. Department of Interior listed a total of 316 species of various types as either threatened or endangered. Over three thousand more were under consideration for listing. Of course, the steady extinction of other species is lamentable. The mortality of species is far sadder an occasion than the death of individuals. Beauty passes from the world, never to be replicated. Nevertheless, the pattern is a common one that is built into the evolutionary process. Humanity masters the entire planet, just as other mammals at one time dominated a valley or a plateau. Inevitably, the human population will expand, taking more and more of the available resources and denying them to other animals. Their death rate will continue to rise while the human population will continue to grow.

We are by nature an expansionary animal. After originating in Africa, we moved into every continent thousands of years ago, surmounting tropical and arctic barriers and spreading over forests, mountains, plains, deserts, and jungles. We are expansionary both because we adapt culturally and because we have turned the juvenile characteristic of exploration into an adult profession. All animals adapt to some degree, but they must do so by the slow process of biological evolution. Only mankind has learned how to invent and apply technology to survive. Of course, all animals, especially juveniles, explore their environment to find food and living space. But only in our species have adults made a career of it. Hence, Captain James Cook, the great English explorer, could declare that he was "employed as a discoverer." The modern world has its astronomers and astronauts who are employed in the same way.

As perceived by a remote intelligence scanning the galaxy, the unusual creature inhabiting the third planet of the solar system of a common G-type star in the Milky Way's Sagittarius Arm may be destined to spread life throughout the galaxy. It is young. It is running out of new habitats on its home planet. And it has the necessary talents: an aboriginal urge to explore, extraordinary adaptability, and very high fertility. Earth may be only its seeding place. In the same way that humanity once lost the knowledge of its origin in Africa as it adapted to life on all the continents—a knowledge that was restored only through years of painstaking fieldwork by dedicated paleontologists—in the same way, the human race may someday lose the knowledge of its mother planet as it brings life to the multitudes of other worlds in the galaxy.

Our knowledge of what an extragalactic being would see as it gazed at the other planets in the Milky Way is also clear. Despite the huge number of stars in the galaxy, there may be no other intelligent life in this part of the universe. In the last twenty years, scientists have recognized that mankind sits atop a pinnacle of evolution that is the result of a long sequence of fortuitous events. The probability of another planet duplicating these events to produce a different form of intelligent life is small. Although scientists cannot state unequivocally that mankind is alone in the Milky Way, the weight of current opinion tends to support that conclusion. For example, Carl Sagan in *Cosmos* estimated that at any given time there were only about ten advanced civilizations in the galaxy. However, he admits that the number may be as small as one. Indeed, some of his assumptions appear too optimistic in light of later evidence. A more recent analysis of this question, authored by both a physicist and an astronomer, concludes that "it is entirely possible, perhaps even likely, that we are the only advanced race in the galaxy." However, "we cannot rule out the possibility that there are a few others, although there is no evidence to support this view."

(Note that this conclusion does not concern the number of habitable planets. The same reasoning yields a conservative estimate of over 100 million habitable planets, taking the number of stars in the galaxy at four hundred billion. The transition from bare habitability to the development of intelligent life is an enormous one. A plethora of suitable planets await interstellar settlers.)

The often-cited calendar metaphor concerning the history of life on Earth can help us understand why intelligence is rare. This analogy puts the events of life's history on Earth from the origin of the universe into the context of a single year. The universe begins at midnight, January 1. Although the sun and its planetary system are formed September 11, the Earth does not become habitable until about October 14; and the first living creatures do not leave the sea until December 24, Christmas Eve. Modern man finally appears after 11:57 P.M. on December 31. All recorded history takes place in the last ten seconds before the end of the year. In terms of the analogy, intelligent life has existed on Earth itself for only about three minutes, while the planet has been habitable over two and a half months. If an advanced, extraterrestrial civilization ever had explored our region of the Milky Way, the chance that it would have found intelligent life on Earth was remote. Our history is too brief an episode.

The scientific reasons for the improbability of intelligent life elsewhere in the galaxy fall into two categories. The first concerns the difficulties life itself faces; the second, the obstacles to intelligent life.

Although scientists do not understand what causes life to develop, they do know what prevents it. They believe that life can originate only in a certain kind of solar system. As many as 90 percent of the stars in the Milky Way are binary; that is, one sun orbits another. Such systems probably do not have planets, since the formation of the double suns consumed the available materials. But if they do contain planets, temperature extremes may easily be too great for life to survive. Also, a star must have a mass that is nearly equal to that of the sun to support zones that are continuously habitable, and only a small percentage of all stars are in this category. But even if a single-sun system has such zones, there may be no planets in them. Had Earth been more than 1 percent more distant from the sun, or 5 percent nearer, life would have been impossible. In addition, each planet must not be too large, or its gravity will be too great for life; and it cannot be too small, or it will not hold an atmosphere. Moreover, a planet may meet all other conditions, but its orbit and rotation may cause temperature extremes unsuitable for life. Finally, no one really knows what caused life. A planet may have the correct sun, the correct orbit, the correct mass, the correct rotation, and still be desolate.

There are also numerous barriers to the development of intelligence. Many favorable evolutionary events occurred before mankind appeared as the first form of intelligent life on Earth. At least half of the history of life on this planet concerns one-celled organisms; more than 3 billion years passed before more complex organisms developed. Biologists have speculated that appreciable and variable tides, causing tidal pools, were necessary for the movement of life to land. But our large Moon causes Earth's tides, and none of the other forty moons in the solar system is as large compared to its planet as the Moon is to Earth, with the possible exception of Pluto's. Dinosaurs dominated life for over 200 million years, two hundred times longer than humanlike creatures have existed, and disappeared for reasons that are still not clear. Did a freakish "death star" or some other improbable event kill the dinosaurs, as scientists are now speculating? And if the dinosaurs had not disappeared, would mankind still have evolved? Probably not. When we examine the evolution of humanity itself, we see that chance played a key role. Man's brain development might not have accelerated had it not been for the rapid succession of ice ages, and the first primates might never have evolved into upright walkers if the tropical forests where they lived had not thinned and become grasslands. The evolution of humanity consisted of many fortuitous circumstances such as these.

Our conception of our place in the universe has now come full circle. For centuries after the fall of Rome, Christian theology taught that the universe was created solely for humankind. The astronomers of the day thought that Earth, at the center of the universe, was the dwelling place of life, while the planets, the stars, and the sun revolved around it. In the last five centuries, science destroyed this view and replaced it with what has been called the "assumption of mediocrity"—the belief that there is nothing special about the solar system or the planet Earth or the life upon it. This premise led to the theory in the early 1960's that the galaxy is populated by many highly developed technological civilizations.

In the last two decades, almost everything that scientists have learned that is relevant to this question has discredited the assumption of mediocrity. Earlier generations of scientists demonstrated that we live in a universe of unimaginable size and age. The work of this generation is making us realize that our galaxy may hold no other creature like us. It has become increasingly apparent that Earth and its inhabitants *are* unusual in the galaxy. This planet orbits a single star of the right mass to have habitable zones; it has moved for the billions of years necessary to produce life in a narrow band where water in its natural state neither boils nor freezes; it has a large moon that causes variable tides and tidal pools; the tilt of Earth's axis is just enough for periodic changes in the climate to occur; and finally, despite enormous odds, intelligent life has flourished here. Although the probability of a planet possessing any one of these characteristics is not impossibly small, the probability that all of them will be present at the same time is. We have reclaimed the idea that humanity has a special place in creation.

Because we are alone in the Milky Way, it is profoundly ours. Other galaxies are too distant to be our concern now; let them be the province of other intelligences from other galaxies. But our galaxy, our Milky Way, offers possibilities for the cause of terrestrial life that will be sufficient for millenia. We are embedded in the galaxy the way a queen bee is in her new and empty hive. The Milky Way is humanity's home, its domain, its kingdom. It is as if all life on Earth were confined to a small, lush tropical island that overflowed with different varieties of plant and animal life while the rest of Earth was virgin land, awaiting the impress of life. The Milky Way's inhabitable worlds await us now. We are destined to be their discoverers, explorers, and settlers. We are the lords of the galaxy.

If humanity is the sole form of intelligent life in the Milky Way—as the available evidence indicates that we are—then our responsibility is grave. We are then the center of life in the galaxy. It makes no difference whether our planet and its solar system lie in the Milky Way's center or on its edge; and it makes no difference if a billion stars outclass our sun: Wherever humanity lives, there live the heart and hope of the galaxy. As the galaxy's only technological civilization, we bear a heavy burden. If life is to continue to grow, if it is to become secure beyond destruction, the decision is ours. If either through natural courses or through some horrible mischance, we disappear from Earth or are severely crippled before we have become spacefaring, there will be an emptiness in the Milky Way far beyond the solar system. The galaxy will remain an uncultivated and barren desert instead of becoming a vast community of life.

THE PROMISE OF SPACE

Thomas Jefferson was a champion of civil liberties and the methods of popular government that enable people to rule themselves. "More than any other man," wrote the distinguished historian Carl Becker, "we think of Jefferson as having formulated the fundamental principles of American democracy." Disliking strong governments and large organizations, he fought for freedom from the British crown, freedom from church control, freedom from a landed aristocracy, and freedom from great inequalities of wealth.

The great aim of Jefferson's career was to give individual men a wider liberty.

Jefferson hoped that the nearly uninhabited continent the new American nation occupied would provide land sufficient for communities of small landholders "to the hundredth and thousandth generation." He believed that such people were the best guardians of America's liberties because they were independent, self-reliant, and individualistic. "They are the most vigorous, the most independent, the most virtuous, and they are tied to their country, and wedded to its liberty and interest by the most lasting bonds," he wrote.

As president from 1800–1808, Jefferson acted boldly to support his views. Without authorization by the Constitution or even the consent of Congress, he doubled the size of the United States by purchasing the Louisiana territory from Napoleon Bonaparte's France. Overnight, the nation grew by more than one million square miles and avoided war over New Orleans with the most powerful country of the day. With one stroke, Jefferson had won peace and secured the land America needed for its growing population. Jefferson also sent out Meriwether Lewis and William Clark—two young Virginians with much frontier experience—on an expedition to explore the unknown regions south of the Canadian border. The three-year, highly-successful Lewis and Clark Expedition accomplished its two objectives of conducting scientific inquiry and opening the Northwest to trade.

The space movement has revived the ideas of Jeffersonian democracy as a realistic alternative in American political life. The expansion of the United States into the new frontiers of space will invigorate the political ideals of civil liberty and self-government that Jefferson loved by increasing American territories on the scale of the Louisiana Purchase. The lands of space are truly sufficient for communities of small landholders "to the hundredth and thousandth generation." We can also expect modern frontiersmen to duplicate many times the bravery and daring of the Lewis and Clark Expedition by exploring the unknown regions of our own era.

Expansion into space does not mean abandoning Earth. Those who cherish life on Earth are a fundamental part of the effort to build a solar-systemwide civilization. The mother planet will remain the source and foundation of life for centuries. Indeed, if *Homo sapiens* is the only intelligent species in the Milky Way, Earth is the precious treasurehouse of the galaxy itself. She must be protected at all costs, or all is lost. The passion and commitment of those who have dedicated themselves to preserving Earth are the foundation of the movement to settle the solar system. Moreover, the space movement may be able to render help in return. If its highest hopes prove true, it can provide priceless services in the struggle to save Earth.

The space movement includes us all. Mass migration from Earth will mean that the economic, political, and educational institutions of society believe that a high return will justify a risky investment of the nation's resources; it will mean that individuals and families are willing to forgo the comforts of terrestrial life for the hardships of a frontier; and finally, it means that we all believe that movement into space offers some possibility of relief from the global crises oppressing us. Space travel will be valueless if it is a cowardly flight from our real duties. The settlement of space requires that private dreams and public responsibilities be satisfied in a civilization with a reach for the stars and an unshakable foundation on Earth.

The Theory of Space Settlement

In 1860, half of the United States was a wilderness. This country, mostly west of Missouri, was a land of Indians and vast herds of buffalo. The Sioux, the Blackfoot, the Crow, the Cheyenne, the Comanche, and the Apache roamed at will, living off the buffalo, which provided everything from food to fuel. Thirty years later, the population of this region was between 5 million and 6 million. The Indians had been defeated, and the buffalo wiped out. The land was crisscrossed by transcontinental railroads, divided among sheepmen, cattlemen, and farmers, and dotted with numerous towns and cities. The American frontier was gone.

During these same years, over 8 million immigrants poured into the United States from Great Britain, Ireland, Germany, and Scandinavia. These people were the hardy folk who settled the West. Their lives on the frontier were hard, dangerous, and sometimes bitterly disappointing. The men worked longer for more meager rewards than farmers in the Ohio and Mississippi valleys; and the women led lives of drudgery and loneliness in homes that were only dugouts or sod cabins in the beginning. Even the towns could be drab and isolated, set in monotonous plains, lacking the trees and bushes and flowers of the settled East. In these settlements, the arrival of the train with newspapers and mail was the most important daily event.

What brought these millions to the United States? Why did they choose to uproot their lives from their old homes and transplant them here? They came because America was the Promised Land. They dreamed of better lives for themselves and their children. Some fled religious persecution; others fled wars and compulsory military service. Some longed for a more democratic society; others came to escape a miserable poverty and to have the hope of getting rich. They did not fear to undertake the great adventure of immigration to realize their dreams.

The settlers of the solar system must also have compelling dreams if they are to be willing to give up the familiar comforts of Earth. The settlement of space, the humanization of the solar system, and the large-scale movement of humanity into the galaxy will only occur if millions of men and women believe that their lives will improve by moving off-planet. Space must be perceived as much more than a desert or a new Antarctica; it must be a place where the old and the young, men and women, children and adults, can lead lives they are happy to have chosen. They will not be moved to choose space by promises of profits to businessmen, power to politicians, or discoveries to scientists. They need something more.

Political movements urging migration into new regions need leaders who are both visionary and practical. As the physicist-philosopher Freeman Dyson noted in *Disturbing the Universe,* these leaders must believe "passionately in the ability of ordinary men and women to go into a wilderness and build a society better than the one they left behind." But if they hope to found settlements that survive, they must also have a clear grasp of the everyday world of people, finance, and technology in which dreams gradually become real. Such a leader was William Bradford, who led the Pilgrims to America in the *Mayflower* in 1620 and who served as governor of Plymouth Colony for thirty years. Bradford established traditions of self-government that set the pattern for America's political development. Another such leader was Brigham Young, who led a party of nearly two thousand Mormons in over six hundred wagons to Utah across the harsh midwestern plains in 1847. Young was an ironfisted administrator who strongly supported education and the theater. He ruled the people of Utah until his death in 1877.

Even if he never leads a trek off this planet, Dr. Gerard K. O'Neill will be remembered as a major leader in the movement to make mankind a spacefaring race. O'Neill, a professor emeritus at Princeton University, first became prominent in the 1950's as the inventor of the storage-ring principle for smashing particle beams that forms the basis of most research in high-energy physics. The publication of his book on space settlement, *The High Frontier,* in 1977, established him as a national space leader. *The High Frontier* was translated into over a dozen languages and remains the most influential book written on space. NASA funded studies of its ideas, and both houses of Congress held hearings featuring O'Neill as the principal witness. *The High Frontier* also sparked the formation of a citizens' group that boasted thousands of members at its peak. O'Neill has continued to be active in the eighties. He founded a satellite-navigation company in 1982 that has attracted more venture capital than almost any other space venture. And in 1985, he was appointed by President Reagan to the National Commission on Space. The commission's report in 1986 endorsed O'Neill's ideas. O'Neill's thoughts on the settlement of space form a large part of the remainder of this chapter.

The genius of Dr. G. K. O'Neill lies in his demonstrating how we can build appealing human habitats off Earth. O'Neill led the effort of the scientific community that showed us how we can assemble cities off-planet using existing technology and attract millions of people into space. These new communities will be attractive because they will contain everything that we love about the natural and social worlds that have nurtured humanity on Earth. The dream that O'Neill offers settlers of the solar system is identical to the one that has always animated immigrants seeking better lives for themselves and their children. It is a dream of economic security and political freedom. It is mixed with the excitement of a new life on a frontier. It is a late twentieth-century reformulation of the political idealism of Thomas Jefferson, who had the same vision of free peoples settling a wilderness. "O'Neill and I have a dream," Freeman Dyson has written, "that one day there will be a free expansion of small groups of private citizens all over the solar system and beyond." O'Neill and his colleagues have shown us how we can build new continents in space that are as suitable for human habitation as any lands on Earth. Life in space can be humane in the deepest sense. Our species will thrive there.

O'Neill's contribution to the humanization of space is not limited to this signal achievement, as important as it is. He has gone further. By showing us why individuals and families in large numbers will choose to migrate into space, he has given us a panoramic theory of the genesis and expansion of extraterrestrial life in the solar system. He has given us a story of how humanity can metamorphose into a spacefaring race and humanize space by building a solar-systemwide civilization. Of course, because of its ambitious scope, this vision could be mistaken in many ways. Nevertheless, the many years over which it has been developed, the critical scrutiny it has received, and the elaborate detail of its planning underscore the realism with which it has been offered to us.

O'Neill envisions the solar system populated by multitudes of spherical cities located in open space that draw their energy from the sun and supply Earth with valuable products. These cities will not be metallic and plastic space stations but Earthlike environments in which the best features of terrestrial life have been duplicated. They will range in size from structures housing a few families to monoliths for millions half the size of Switzerland.

A typical city will consist of a collection of small villages separated by forests and parks, nestled in valleys and on hills. Its inhabitants will live under blue skies with natural sunshine, normal gravity, and the usual cycle of day and night. Living in spacious homes, they will enjoy flowing rivers, sandy beaches, and grassy fields. Flowers will be abundant, fresh vegetables commonplace, and butterflies numerous. Light rain can be timed to fall before sunrise daily. Residents will be young, old, families, singles, children, and seniors; they will form a balanced and complete community. Animals such as birds, squirrels, deer, dogs, and cats will remind them of Earth. Since parts of the city will have little gravity, space sports such as human-powered flight will become popular. O'Neill writes in *The High Frontier,* "The general impression is of a village with greenery, trees and luxuriant flowers."

In contrast to NASA's space program—which constantly requires new technology—O'Neill's space cities are based on the principles of ordinary civil engineering. Advanced technology is not necessary. In fact, O'Neill's plans do not even need new rockets to transport construction materials into space, since the most economical source of supply will be the Moon. Because of the *Apollo* missions, we know that the Moon can furnish all but a small part of the materials needed to build space cities. Nor is their scale unusually large. O'Neill estimates that ten thousand people will be able to live comfortably in the first habitat. Yet this "island in space" will have about the same mass as a seagoing supertanker and be composed of materials that will leave an excavation on the Moon of only five yards deep and two hundred yards square.

An important characteristic of space cities will be their mobility. Most may choose to remain in one location, near Earth, for example. But others may use the Sun's energy to embark on slow voyages to other parts of the solar system for scientific or commercial purposes. According to careful calculation, space cities will be able to roam out as far as ten times the orbit of Pluto—the outermost planet—and still draw sufficient energy from the sun to allow their inhabitants to carry out their normal lives. The entire solar system may one day also become a base for interstellar migration. Connected to independent power sources, space cities will be able to move on multigenerational journeys to nearby stars.

The continuous and thorough engineering review of O'Neill's concepts for over ten years distinguishes them sharply from the ideas of ordinary futurologists. "The engineering aspects of [the space city idea]," testified the astronomer Carl Sagan to the House of Representatives, "are perfectly well worked out. . . . It is practical." Since O'Neill is by training a scientist, he subjected his ideas from the beginning to the exacting empirical standards of his profession. He sought to publicize his vision as widely as possible, invited criticism, and scrupulously answered objections. He made dozens of speeches, appeared in many periodicals, and presented his plans to both Senate and House subcommittees. Of course, the act of publishing a hypothesis and allowing it to stand on its own merits is a standard practice within the scientific tradition. The manner in which he handled correspondence reveals his high standards:

> The typical letter was thoughtful, lengthy and represented a considerable input of effort and study on the part of the writer. For those reasons, the mail could not be answered in a careless or standardized fashion: a careful, thoughtful letter demanded an answer of the same quality. For more than a year I struggled to keep up with the mail myself, but the rate increased continually, the quality remained high and . . . the burden became too great. Since that time many of the letters and requests for information have been answered by a group of volunteers, each expert in a particular research area. Some letters, especially thoughtful and helpful, I must deal with myself. . . .

A passage in *The High Frontier* illustrates the critical process through which the concept of space settlement has passed. O'Neill states in the section on the structure of the space habitats that he does not need to discuss in detail the metal shells that contain the forces of atmosphere and rotation because "A number of engineers in several government and private industrial laboratories have checked the relevant calculations." O'Neill's choice of the first publication to disseminate his views is also illustrative. He chose the professional journal *Physics Today*, thereby inviting criticism from some fifteen thousand practicing physicists. Many attempted to find flaws in the arguments, to discover numerical errors, and to point out faulty assumptions. Standing before the bar of his peers, Dr. O'Neill devoted a great deal of time answering these critiques

in detail. Some of his responses ran to twenty pages of closely
reasoned arguments and calculations.

O'Neill's ideas have been published and critiqued so widely
in technical journals, research groups, and university conferences
that they are more truly viewed as the product of the science and
engineering communities as a whole than as the work of one man.
In 1976, NASA assembled a very high-level team of aerospace
professionals to study three key technical questions. Each partic-
ipant arrived at the study with documents and calculations result-
ing from many years of practical experience. They were assisted
by an excellent group of students and top-level specialists brought
in as consultants. Another example of the community effort involved in elaborating O'Neill's ideas is the conferences that have
been held at Princeton University since the mid-1970's on space
cities. The first drew over 150 participants who presented papers
on technical subjects and debated controversial points. The second
conference was longer, better attended, and cosponsored by Princeton and the American Institute of Aeronautics and Astronautics
(AIAA), the professional society of the aerospace field. It was also
financially supported by grants from NASA and the National Science Foundation. Researchers presented over thirty papers. Because of the growth of interest, attendance at the third conference
was limited by invitation. By 1987, there had been eight such
conferences.

The numerous research groups that sprang up at several universities in the mid-1970's provide further evidence of the involvement of the scientific establishment. By the end of 1975, active
groups of students and faculty at universities such as MIT and
Polytechnic Institute of New York had begun research on a vol-
unteer basis. By 1976, the research group at Princeton had grown
into a close-knit organization with nine members, half of whom
worked as volunteers. From this group the Space Studies Institute
(SSI) evolved. SSI is today supported by donations from thousands

of private citizens; its sole purpose is research on subjects related to the settlement of space.

O'Neill failed to attract the support he needed to make his ideas an official part of America's space effort, perhaps because he proposed them at the wrong time. When he started to receive national attention in the mid-1970's, the centerpiece of the U.S. space program was the reusable space shuttle. At that time, the space community's support with Congress and the public had fallen so low that insufficient funding forced NASA to reduce drastically the shuttle's scope from the original design. Year after year, NASA's budgets fell, compelling it to disband the triumphant teams of engineers, managers, and contractors that had made the Moon landings possible. Except for the brilliant results of the planetary probes, the 1970's were dismal times for America's space program. O'Neill had little chance of winning votes for a costly new program when existing projects suffered from inadequate funding.

The O'Neill program will be expensive whenever it is undertaken. A NASA study estimated the cost of building the initial habitat for a population of ten thousand to be about four times the cost of the entire *Apollo* project. However, this estimate includes as major components the costs of establishing a Moon base and of boosting into orbit ten thousand people at space-shuttle prices. Both these costs can fall dramatically. As we have seen, the nations of Earth have many reasons to become space powers. They will one day crowd near-Earth space with their manned stations in the same way that communications satellites now crowd the available slots in geosynchronous orbit. Military and commercial activities will lead to a Moon base, and the development of a cheap propulsion system will accelerate the decline in costs even more. O'Neill made his proposal at a time when virtually no infrastructure existed in space. When an elaborate one is in place for other reasons (such as for business or defense), the migration from Earth can begin in earnest.

O'Neill's principal accomplishment is technical. Not only has he demonstrated that the concept of cities in free space is feasible, he has also shown that much of man's natural environment can migrate off-planet with him. We can build habitats in space that can rival Earth in their congeniality to humankind. The report of the National Commission on Space, *Pioneering the Space Frontier,* in 1986, states unequivocally that "conceptual engineering studies confirm that attractive settlements of this kind [with Earthlike characteristics] can be built in orbital space. Such habitats could accommodate thousands of people." Indeed, O'Neill's appointment to the commission indicates the position of leadership and respect that his work has won for him in public life.

In spite of many questions yet to be answered and some ideas still being debated, Gerard K. O'Neill's intellectual achievement is enormous. He and his followers have charted the first path with realistic calculations that mankind can follow in its journey to become as spacefaring as fish are aquatic. He has given us a clear vision of a solar-systemwide civilization in which a large-scale expansion of *Homo sapiens* has humanized space. Furthermore, he has demonstrated how life off-planet can attract the millions of people needed for the new frontiers: They will come for the chance for a better life for themselves and their children; they will come because of the beautiful new lands; they will come for economic security free from corporate and government restrictions; they will come for wealth and adventure. They will not be fleeing Earth as exiles. O'Neill has shown us how mankind can escape the bonds of its home planet, continue to grow in numbers and prosperity, and begin to bring life to the rest of the galaxy.

THE HOPE OF SPACE

The hope the space community offers the world is the possibility of relief from the worsening problems of overpopulation, nuclear war, resource shortages, and ecological disaster. Most advocates of space development recognize the speculative nature of this hope. O'Neill's plans for space cities are solidly grounded in ordinary civil engineering. Believing that expansion into space will rescue humanity from its plight requires political and sociological predictions that are less certain. Nevertheless, the glimmer is there. It exists.

We cannot forget how to make and use nuclear weapons, but we can reduce the deadliness of their effects. Large-scale expansion into space will allow us to transcend the danger of nuclear war annihilating humanity. As mankind becomes dispersed throughout the solar system in numerous communities, the danger of destroying all life with a single exchange of missiles diminishes. In the 1980's, Soviet missiles can explode on American cities within minutes after launch. Defense is impossible; and because the Earth is a closed system, fallout and aftereffects such as a nuclear winter will be equally deadly. But if the solar system is filled with space communities to beyond the orbit of Pluto, single-strike attacks would be impossible to coordinate. Even a nuclear-powered missile would take months to reach its target, and the destruction of one city would not imply the destruction of others. The awful possibility that a single aggressor could murder Earth's life would be gone.

The settlement of space will contribute to peace in another important way. If our aggressive and energetic species remains focused on this planet, Earth will become a pressure cooker with a mounting temperature that will someday explode. If, on the contrary, we redirect our attention outward, space becomes an outlet for our often-violent passions and ambitions. No longer are more and more fighting for less and less; the terms of the struggle will have changed.

Space mines will reverse the current trends toward scarcity. All the resources humanity needs to support a growing population for millenia are available in space. The mineral and energy wealth in the inner solar system, the asteroids, and the outer planets is sufficient to sustain a human population many thousands of times larger than it is now. If we do develop an inexpensive propulsion system and become spacefaring, we will use these resources to build an interplanetary civilization and relieve any shortages on Earth.

The hope that migration into space can relieve the world's population problem may be unfounded. All writers on this subject strongly emphasize the speculative nature of the question. Few attempt to make a prediction; most are only interested in establishing a possibility. O'Neill, for example, is careful to distinguish his speculations on population from the rest of his work. He states in *The High Frontier* that he is not offering prophecy; he is only calculating what will be technically possible with the conservative assumptions that he has been using:

> When we consider this possibility of reducing the population of Earth by emigration, it is important to distinguish possibility from prophecy. . . . the combination of technique with natural growth in capability would have the *power* to permit such emigration. Whether or not large scale emigration will occur will depend on how badly it is needed, and on how attractive the space communities become.

O'Neill starts his reasoning after mankind has acquired the skills to construct space habitats routinely. He believes that the pioneering days will be passed when sixteen communities have been built with a population of about 160,000. Assuming productivity grows slowly, he writes, "it would require only thirty years from the completion date of the first community before new lands would be increasing more than fast enough to cope with [an increase in Earth's population of 200 million per year]." An increase of 200 million per year is the expected annual growth of a world population of 10 billion (which reached 5 billion in the mid 1980's). By comparison, the number of passengers flown worldwide by commercial airlines rose to 320 million in 1984 from a few thousand sixty years before. A commercial space industry could be transporting millions of passengers a few decades from now.

The resolution of Earth's ecological problems is the most remote possibility offered by expansion into space. *If* we learn how to build new lands in space accommodating much more than the annual population increase, and *if* life in these new communities is more attractive than life on Earth, *then* Earth's population may fall. More and more areas will revert to wilderness, with bird and animal populations increasing. Mankind's home planet will become a residential parkland for a relatively small and affluent human population.

By making fanciful forecasts about such results of space migration, space activists have often eroded public confidence in the more reliable claims of the space community. No more damage need be done. Nevertheless, the hopes of resolving the world's major problems may not be as distant and as vaporous as they appear. Consider these thoughts: The likelihood that engineers will design an inexpensive means of transporting masses of people into space is high. Remember how few years separated the Wright brothers' flight from a commercial air industry flying hundreds of millions of passengers annually. Second, O'Neill has shown us that the natural milieu that has nurtured us can be recreated off this planet. Yes, in the 1980's, with virtually no space infrastructure and with an exorbitantly expensive space shuttle, the cost is prohibitively high; but these circumstances may vanish. Finally, why should the march that life began billions of years ago at the bottom of the sea halt at the edge of the atmosphere? Man has the ability to become a spacefaring creature, and he has been preparing himself for the leap off this planet for decades. If he succeeds, life will not content itself with scattering a few spores into nearby space. Earth will become the fount of life for the galaxy. It will flood the stars with life the way oceanic life once overflowed onto all the earth's barren land.

THE GREAT OPPORTUNITY

The solar system is a natural habitat for humanity. It is rich in resources, traversable in reasonable periods of time, and separated from the rest of the galaxy by huge distances. In its suitability for human occupation, it resembles the sites of mankind's first civilizations. The great river valleys—the Nile, the Indus, the Yangtze and the Tigris-Euphrates—were also natural habitats, because they also were rich in resources, traversable by the transportation methods of the day, and sealed off from the rest of the world by vast, unknown lands. Until the late twentieth century, Earth's atmosphere walled us off from the solar system, and we built a global habitat by learning how to travel rapidly anywhere within it to get what we wanted. We are now ready to move on.

We must begin to build a solar-systemwide civilization for the sake of humankind. Spreading life from the Sun to Pluto will free us from the danger that madmen or merely erring leaders will extinguish humanity from the galaxy through nuclear war. A human community extending the immense distances from the Sun to Pluto—the outermost planet—would be beyond the possibility of destruction by even the most determined militarist. We can also hope that a truly spacefaring humanity equipped with an inexpensive mass-space-transportation system will reduce Earth's overpopulation, replenish dwindling mineral supplies, and relieve ecological problems.

We must also move into space for the sake of our children and their children. The wealth of the solar system is as virginal as North America was in the seventeenth century, and we can take much that we love about Earth's natural world with us. Life in new settlements can be less regulated, freer, more affluent, and less endangered by nuclear war. We need not worry that we will consume the available resources, for there is enough for multitudes of generations.

The foundation for mankind's expansion into space will be the scientific, political, and commercial advantages that will accrue to all space powers. Any country that ignores the potential of space for national aggrandizement in the twentieth century will surely be subordinate to those that did not in the twenty-first century. As has always been the case, the fuel that will drive men daily toward new frontiers will be their desire for fame, power, and profits. The momentum to move off this planet is strong because space can satisfy these desires in many ways.

The conclusion that we have a moral imperative to make Earth part of a much larger civilization is not utopian in the context of the alternatives. Our intellectual leaders have been preoccupied with the problems of a nuclear world since the first atomic explosions at Hiroshima and Nagasaki in 1945. Yet, in over forty years, they have offered us little on which to base hopes for a positive future.

Jonathan Schell wrote a brilliant, influential, and widely praised book in 1982 entitled *The Fate of the Earth*. Mr. Schell took as his subject the great truth of our time that nuclear war may lead to the extinction of the human species. In a field in which thousands of books have been written, his success was extraordinary. Establishment leaders rushed to give him well-deserved tribute, and major publications—*The New York Times*, the *Washington Post*, *Newsweek*, the *Christian Science Monitor*, *Vogue*, the *Baltimore Sun*, and the *Chicago Sun-Times*—lavished praise. One critic believes that Schell has written one of those rare books that make political history, comparing it to Harriet Beecher Stowe's *Uncle Tom's Cabin*, Émile Zola's *J'Accuse*, and John Maynard Keynes's *Economic Consequences of the Peace*.

After a lucid exposition of the world's nuclear predicament, Schell discusses what mankind must do to avoid extinction. According to his analysis, we must take three actions to preserve humanity: total disarmament of both nuclear and conventional weapons, relinquishment of sovereignty by all the nations of the world, and formation of a new political system for the peaceful settlement of international disputes. Schell believes that politics must be revolutionized to create a political means by which the world can make decisions sovereign states now make by war. "For today the only way to achieve genuine national defense for any nation is for all nations to give up violence together."

However one may judge Schell's proposals, they are representative of the solutions to the nuclear dilemma that our leaders have offered us. Whether they propose a nuclear freeze or the abandonment of the rights of sovereignty by all the nations of the world, they ask us to believe that the historical patterns of international behavior will change in a fundamental way. In response to the dire and unprecedented threat of nuclear war, we are told, men and nations must behave differently from the ways they have behaved for millenia.

The project to make humanity spacefaring and to build a community that spans the solar system has its own uncertainties. Nevertheless, it is a course of action that has many biological and historical precedents and is rooted in trends in technology that are decades old. Movement by a successful species into an adjacent habitat is a commonplace phenomenon in nature; world history contains many examples of large migrations of peoples to new lands for the sake of better lives; and since the Soviets launched *Sputnik* in 1956, man's activity in space has steadily increased. Calling for an expansion of the human biosphere is not the radical innovation that conventional solutions to the nuclear problem appear to be. It is a natural and gradual evolution from our present way of life that will allow mankind to transcend the danger of extinction through nuclear war.

As citizens of the world concerned with our loved ones and the future of our species, we must encourage the work of both those seeking political solutions on Earth and those seeking technological solutions in space. Our job is to maximize our chances for survival by neglecting no possibility. Our circumstances are so grave and so problematic, with so small a margin for error, that we cannot afford the loss of any grounds for hope.

Let us therefore go forward supporting both space-based and earth-based solutions to the problems of our age. Stifling the creativity or commitment of either party may be suicidally short-sighted. Both have their strengths and their weaknesses, and they are by no means antagonistic to each other. Both are suffused with a high idealism and a passionate concern for humanity's future. Both are based on ideas developed by long and laborious thought, and both are led by deeply honorable men of acute and comprehensive intelligence. And both—despite the perhaps more practical approach of the scientific community—have distant visions of faint possibilities that will never become real except through the moral heroism of us all.

After millions of years of evolution, countless experiments, and countless failures, one very new life-form has mastered Earth in a startlingly short period of time. This successful species has unusual powers and is expanding into every available habitat on the planet, thereby bringing to extinction many less fit forms of life. It is encountering little resistance and its population is increasing geometrically. Earth bursts with its kind.

Fortunately, at the very geological moment that this species is finding its home planet too small, it is acquiring the means to leave it. It is becoming spacefaring. It has discovered that its planet is only a very small island in a galactic sea containing a large number of planets as suitable for life as its own. It is beginning to realize that it is perhaps the dominant form of life not only in the solar system but in the entire galaxy. Although life is surely present somewhere, the long series of events that produced intelligent life on Earth is unlikely to have been duplicated.

Humanity has the opportunity to become the bearer of life to the galaxy of stars it inhabits. If it can avoid the catastrophes of war, overpopulation, and ecological disaster that threaten its home, this race can bring the light of life to multitudes of planets and become in deed the leading species in the Milky Way. Earth floats in a void surrounded by possibilities so great that they dwarf all that life has wrought to date on this planet. In the same way that the variety of life on land has become orders of magnitudes greater than the vast population in Earth's ancient oceans, so do the prospects for life in space exceed those on Earth. To be content with life's present boundaries on this planet would be as myopic as contentment with their lot would have been for the first city-dwellers ten thousand years ago. We shall survive; we must go; we shall triumph.

When we look out upon the numberless stars of the Milky Way and feel our insignificance before them, we recapture emotions the first men must have felt. Gathered in tiny, defenseless bands, surrounded by an uncountable variety of hostile animals and by an extent of land and sky that seemed without end—how could they have felt any differently than we do today, as we think of our galaxy? If the thought of humanity peopling the Milky Way seems unreal, would it not have seemed equally improbable to our ancestors that their descendants would one day cover the earth, gradually gaining mastery over all other life? So too must we measure our own doubts when we look at the stars and wonder whether our powers are a match to theirs.

CHAPTER FIVE

A Heroic Age

We are entering an age in which great deeds are possible. In the next fifty years, men and women alive today may perform acts that will survive in the memory of our race as long as we call ourselves human. A heroic age is coming. Fresh heroes will open doors that will lead to vast vistas and embark on enterprises that will endure for lifetimes. Conferring benefits on us all, their actions will be the stuff from which myths and legends spring. The historian William McNeill declares that people of the future will look back upon our era "as a truly extraordinary Golden Age, when everything was open, everything was up for grabs." Only the weak and timorous, he continues, can "regret being alive in a time when so many avenues lie open and so much remains to be done."

The great deeds to be done are those acts that make human-kind spacefaring by expanding the human habitat over the entire solar system. This is the time to begin to build a civilization that spans the planets, from Mercury to Pluto. These are the days that we begin to build a civilization whose citizens will contemplate with satisfaction a system of worlds providing comfortable lives for humanity's billions; who will regard the clear skies, clean rains, and green fields of Earth as an ecological treasurehouse; and who will feel freed by the knowledge that war cannot destroy all life. The people inhabiting this broader and richer community will be a breed superior to us, who now live on this cramped world with its noisome air and fragile peace.

Among all the acts that will make an interplanetary civilization possible, none is as important as opening up the regions outside this planet so that they are accessible to millions. Spaceflight must become cheap and ordinary. It must become a mass-produced service available to everyone. Our task is *not* to launch space shuttles, build space stations, return to the Moon, or land on Mars. These achievements are important results of our supreme objective—inexpensive spaceflight—from which all the beneficial consequences of a multiworld civilization will follow. When access to space costs little, the great rush will begin and a new age for humanity will have arrived. And those who make this change possible will earn the world's gratitude and history's attention. But nothing can happen while we are confined to this single planet. As long as Earth is our only home, we will continue to pollute, over-populate, and accumulate weaponry. Only when space travel has become as common as air travel will the new era dawn.

To accomplish this task, thinkers must broaden their conception of the human world, engineers must develop the technologies to reduce launch costs, business must build enterprises that provide cheap launch services, and politicians must synchronize America's space program with the nation's political processes. The writers, engineers, businessmen, and politicians who devote themselves to settling space are our true public servants, and those who succeed in their purposes will become the heroes of these times.

The reform of America's space program is the most important task of all. The U.S. space program is failing. The federal government has mismanaged the manned space effort, outraged businessmen and scientists, and acquiesced in a shameful militarization of space. In addition, the USSR will widen its lead and other nations will soon surpass us too, unless we restructure our space program. The American space effort is based on a mistaken understanding of what we can do, and not on our national strengths. We are not a nation that has any talent for using the government to accomplish vast designs over long periods of time. Our fundamental political process is individualism, and a successful space program must be based on this fact.

An interplanetary civilization is inevitable. The day will come, prophesied H. G. Wells at the beginning of this century, when "man will stand upon the Earth as a footstool, and reach out his hands among the stars." We are nearer that day now than we have ever been before. After starting in the wilds, humankind built farms and formed villages that became cities. First we occupied fields, then valleys, then regions, and finally whole continents. We now possess a planet, and have been shooting tendrils into space in increasing numbers for thirty years. The tide runs in only one direction. Let us not stand passively by the flood as it rushes past but shape its courses. We have traditions to maintain and standards to uphold, false directions to avoid and mistakes to correct. Grand opportunities and trying passages await us. The time is now.

INTELLECTUAL LEADERSHIP AND A NEW PUBLIC PHILOSOPHY

The size and perseverance of a voting public that does not identify with either major party indicate that significant political needs remain unsatisfied. For the last fifteen years, the percentage of voters who consider themselves independents has remained at about 30 percent, which is about the same percentage that consider themselves Republicans. The persistence of these moderate-road voters suggests a demand among the electorate for a public philosophy that combines the best features of liberalism and conservatism.

The steady decline in voter participation in the last two decades leads to a similar conclusion. While democracies in other countries (such as Canada and England) regularly achieve high voting levels, voting peaked in the United States in the presidential election of 1960 at 62.8 percent and has fallen almost ever since. It dropped to 53.5 percent in 1976 and to 52.6 percent in 1980, rose half of one percent in 1984, and dropped again in 1988 to 50 percent—the lowest rate since 1924. Certainly, other factors are at work. Increasingly lengthy campaigns have alienated voters too. But part of the disaffection must be due to a dissatisfaction with what is offered.

The phenomena of the independent voter and the nonvoter are symptoms of a more serious issue. The public philosophies that have guided national politics since the Great Depression in the 1930's are inadequate to present circumstances. For the last fifty years, conservatives and liberals have fought one another for the presidency, in the Congress, in the courts, and in the states. Sometimes one side has prevailed, and sometimes the other. Each party may rightfully claim victories that served the national interest. The left was responsible for holding the country together in the Depression and for the landmark civil-rights legislation of the 1960's; the right opened China in the 1970's and brought down inflation and unemployment in the 1980's. But now—on the eve of this generation's political maturity—both conservatism and liberalism have lost much of their appeal. For many of us, neither party offers a vision that explains the present and gives us hope for the future.

The right may have the greater problem. Although its economic policies have many attractive points, its social policies do not. Conservatives lack sympathy for the have-nots and the disadvantaged of our society. The Republican party, for example, is notorious for its neglect of the black vote. Moreover, conservative economic philosophy has never sufficiently explained the role of the government. An active government is not inconsistent with policies that encourage the private sector, as conservative antigovernment rhetoric implies. But the most damaging issue for the right may be character. As evidenced by Watergate and the numerous scandals that plagued the Reagan administration, the Republican party attracts more than its share of men who place private profit above public service. Many people do not trust Republicans. For this reason, registered Republicans have rarely exceeded 30 percent of the electorate in the last thirty years. Pollster George Gallup, Jr., in 1987 reported, "Today, Republicans hold no significant advantage in any major population group."

The problems of the left are complex. On one hand, the liberal sense of community and its belief in an active government have produced the only genuinely heroic political figures of late twentieth-century America—Martin Luther King, Jr., John F. Kennedy, and Robert F. Kennedy. These attitudes enabled the Democrats to retain the allegiance of the majority of voters for fifty years, regularly outpolling the Republicans by one third. On the other hand, many regard traditional liberal altruism as anachronistic in a world in which the United States economy is no longer preeminent. In an era of severe budget constraints, the Democrats cannot deliver on all the promises they make, and they do not have a definition of the public good that permits them to adjudicate the individual claims of their adherents. The liberal public philosophy, declared Robert Reich, an influential liberal author and member of Harvard's Kennedy School of Government, "no longer embodies a story in which most Americans can believe." The Democratic party, he wrote, became by the late 1970's "less the embodiment of a shared vision and more a tangle of narrow appeals from labor unions, teachers, gays, Hispanics, blacks, Jews, the handicapped, the elderly, women."

If it is possible to combine the best of the right and the left, our generation, by virtue of our experiences, may be best equipped for the task. In some ways, our lives mirror the nation's political composition. In the 1960's, we tried the left. Those years were times of public purposes and common goals. We marched for peace; we marched against the military; we marched for civil rights. We worked in the inner cities; and we worked in the Third World. In the 1970's and 1980's, we tried the right. We returned to our private lives and founded numerous groups, therapies, cults, and movements to develop ourselves as individuals. In the present decade, we have made businessmen folk heroes and have twice helped elect Ronald Reagan president. But these times have passed. We—together with the nation—need a new public philosophy to meet the frightening demands of the coming years.

In an important essay, "Toward a New Public Philosophy," published in May 1985, Robert Reich of Harvard University discussed the elements of a new synthesis. This new philosophy must have a central theme to organize and legitimize the claims of its diverse constituencies. It must have a strategic approach for global change. It must define America's place in a changed world. It must seek an enhanced quality of life for all. It must reject the notion that the central struggle of our age is over the division of a fixed quantity of global wealth. It must simplify and reassure. It must prescribe clear lessons. And it must explain the new reality.

The opening of a space frontier satisfies Reich's criteria for a new public philosophy in many ways. But despite the spectacular progress humanity has made in space since the 1950's, widely read writers and thinkers have given little attention to extraterrestrial settlement. Few speak for space. No book treating movement off Earth has ever appeared on the best-seller lists, and no author on space settlement has ever won any prestigious award such as a Pulitzer Prize or a MacArthur Fellowship. Sometimes the neglect is blatant. Allan Nevins and Henry Steele Commager wrote an excellent history of the United States that has been through seven editions since it was first printed in 1942. The most recent edition, which was published in the mid-eighties, covers American history up to the Reagan administration. Yet it ignores the Moon landing, the historic interplanetary probes, and the space shuttle. Moreover, the tone of its authors is decidedly not expansive. They write of the demise in the scandal-racked seventies of the idea of "an infinity of land and resources" and of the idea "of a special destiny and a special mission" for this country.

Our intellectual leaders must provide guidance before we can begin the work of expanding the human domain. Because serious thinking on the significance of space has largely been confined to the scientific and engineering communities, popular misconceptions are commonplace. Many people, for example, believe that space settlements will resemble either claustrophobic submarines or low-gravity hotels, such as Stanley Kubrick depicted in the film *2001*. Both notions are mistaken. More seriously, some believe that opening the space frontier means that we are abandoning Earth and neglecting our responsibilities here. They think that developing this planet's deserts, polar regions, and undersea lands would be more fruitful and less expensive. These ideas are also in error. But perhaps the most damaging popular concept is the widespread conviction that only a highly trained elite will see the grandeur of Earth from space in the next decades. We are capable of far more than this bleak view indicates.

Settling space means that we will enlarge the kind of habitat in which human life has flourished here on Earth. We will become exporters of all forms of life, and creators of the kind of natural world in which humankind can be happy. Space settlements will resemble small midwestern towns. Not only will they have normal gravity and sunshine, but they will also have sufficient territory for flowers, grass, trees, wildlife, and lakes. They will rival the best that surface cities can offer. Life in space may even be more attractive than life on the ground because of the strong sense of community that will surely develop, the multiple opportunities that a frontier always presents, and the new sports that low-gravity playgrounds will make possible. And space settlements with these features are not the fanciful dreams of a science-fiction novelist.

Studies conducted by aerospace firms, universities, and NASA in the 1970's demonstrated that orbiting habitats congenial to people are feasible. The National Commission on Space confirmed this result ten years later. Its 1986 report stated that "attractive settlements of this kind [as described above] can be built in orbital space" and "could accommodate thousands of people."

We can resolve Earth's crises neither by remaining on this planet nor by seeking isolated havens in space. Between these extremes lies the moderate course of building a solar-systemwide civilization that will transcend our troubles. The currents of population, the ubiquitous drive to industrialize, and mankind's propensity to war lead to catastrophe on a closed planet. Nor should any group suppose that it can survive long off-world without the sustaining presence of Earth. The umbilical cords interconnecting us are too numerous to cut. But for a spacefaring species, having a habitat extending the breadth of the solar plain, the possibilities for life are much greater than terrestrial deserts and ice caps could ever provide. We shall find homes for the world's billions, purgation for this planet's ills, and outlets for our aggressions by obeying life's ancient imperative of growth. There is no other way.

The opinion that decades must pass before the space frontier opens to us all is false. We do not have to accept such an unhappy future. Great peoples rise to challenges considered beyond their reach. They are capable of cultural jumps in dire circumstances. The biblical Jews, the ancient Greeks, the mighty Romans, the imperial Chinese, the conquering Muhammadans, the medieval Christians, the extraordinary Europeans, and our own Founding Fathers each had their days of great challenge and high response. But their times have passed. This hour is ours, and it calls forth all that we have within us. We spring from noble lines with a rich heritage and wonderful models. At no other period in human history has there been such an outburst of innovation. Our age rivals Periclean Athens and Renaissance Italy in creativity. In less than seventy years, we went from the Wright brothers' flights of a few feet to a flight to the Moon. We can democratize space if we choose.

We need public intellectuals to correct misinformed judgments such as these and others. Space settlements will be natural worlds filled with Earth's life. Settling the solar system is a centrist and responsible way to resolve the global crises. We can build mass-space-transportation systems soon. Intellectuals can also awaken us to the possibilities before us. The power of space settlement as an alternative lies in its broad effect on human endeavor, from politics and economics to religion and private life. Thinkers and writers can make us aware of our options, foresee difficulties, dispel errors, and ignite our enthusiasm. We need their help. We need intellectual leaders to form a new public philosophy that does not ignore the benefits of an open space frontier.

TECHNOLOGY

Compared to what politicians, businessmen, and intellectuals must do to make space settlement possible, the task of engineers appears less daunting. They have been developing space technology for decades, and momentum exists in their field that is just beginning in the others. In fact, most experts agree that the technology to move off Earth exists now or can be developed in a reasonable period of time. As early as 1979, a NASA report, *Space Resources and Space Settlement,* concluded that a large industrial base could be established in space within this century. Other reports have reached similar conclusions. If the broad development of technology over the last few centuries continues, human civilization will grow beyond this planet and spread throughout the solar system.

We rely on the ingenuity of an army of engineers to perform this essential work. Working in the government and the aerospace corporations, they will design the technology that will make humanity spacefaring, and they deserve honor for their role. Engineering work in large corporate and governmental bureaucracies is, without question, essential to the conquest of space. Almost all that we have accomplished so far is because of them.

Some achievements, however, may lie beyond the reach of the employees of government and aerospace corporations. These organizations are not the only means to technological breakthrough. Individuals and small groups, often without institutional support, depending on their own initiative and devotion, have been responsible for many creative jumps in the history of technology. Perhaps one of the most famous instances occurred in the development of the airplane. Samuel Langley was the secretary of the Smithsonian in Washington, D.C., at the turn of the century, and had the finest engineering minds of the day at his disposal. He received fifty thousand dollars from the U.S. government to build an aircraft, which sank into the Potomac River seconds after takeoff, in October 1903. Orville and Wilbur Wright, on the other hand, were bicycle mechanics who spent less than one thousand dollars to build the biplane that flew at Kitty Hawk a few weeks later.

The history of spaceflight contains similar stories. In Germany in 1927, nine young men formed the Space Travel Society. Before Hitler disbanded it in 1933, the society had worked through the basic engineering of liquid-filled rockets and had sent several flying to heights of one to two kilometers. They had received no help from their government. In recent times, there have been less successful although equally inspirational attempts to advance space technology without the big bureaucracy and big money characteristic of government projects. In 1958, in San Diego, California, a small group of engineers began work on a spaceship intended to carry large payloads cheaply all over the solar system. They wanted to build the twentieth-century equivalent of sailing ships like the

Mayflower, which brought the Pilgrims to New England in 1620. Called Project Orion, the group made considerable progress on a nonchemical propulsion system before losing its funding in 1965.

Despite Orion's failure, the possibilities it represented made a deep impression on one of its members, Freeman Dyson. Dyson is a theoretical physicist who has done work in the fields of nuclear physics, astrophysics, and rocket technology. He has been a professor of physics since 1953 at Princeton's renowned Institute for Advanced Study, where Albert Einstein worked from 1933 until his death in 1955. Dyson, who called the time he spent working on Project Orion "the most exciting and in many ways the happiest" of his scientific life, believes that Orion is an example of the kind of independent thinking in propulsion systems that suffocates in government projects:

> The history of the exploration of space since 1958 has been the history of the professionals with their chemical rockets. The professionals have never been willing to give a fair chance to radically new ideas. Orion is dead and I bear them no grudge for that. Orion was given a fair chance and failed. *But there have been several other radical schemes that came later, schemes better than Orion, schemes that could do everything Orion could do and more* [emphasis added]. . . . None of these newer schemes has been given the chance that was given to Orion, to prove itself in fair competition with chemical rockets. Never since 1959 have the inventors of new kinds of spaceships been encouraged to try out their ideas with flying models as we did. . . .

Huge corporate and governmental organizations lay behind the progress humanity has made in space to date. Multilayered bureaucracies employing tens of thousands, using hundreds of subcontractors, and costing billions of dollars are responsible for both the American and Russian space programs. We shall not dispense with their services soon. In the near future, the development of space will remain the province of governments and large corporations. But there is still a place for the individual with a dream and the small group devoted to an unconventional idea. The course of human affairs is cyclical, and the time of monumental organizations will pass again. Dyson thinks that the belief that space projects must cost billions of dollars is wrong. "A small group of people with daring and imagination," he says, could design a space ship that would be cheaper and much more capable than anything available now. "I believe," he declares, "the road that will take mankind to the stars is a lonelier road, the road of [Russian rocket pioneer Konstantin] Tsiolkovsky, of Orville and Wilbur Wright, of [American rocket pioneer] Robert Goddard and the men of the [German] Space Travel Society, men whose visions no governmental project could encompass."

BUSINESS

When gold was discovered in California in 1848, the world rushed in. Most of the population of northern California stampeded to the mines. San Francisco lost two thirds of its inhabitants, and many crews of the ships in its harbor abandoned their crafts. The commander of the U.S. Pacific squadron believed that it would be impossible to maintain any military establishment for years. "To send troops out here would be needless," he wrote to the secretary of the navy, "for they would immediately desert." Hundreds, then thousands, then tens of thousands, of men poured into California from Mexico, Hawaii, Oregon, China, and the Eastern Seaboard. In one year, the discovery of gold transformed California from an obscure frontier in the American West to international prominence.

Before 1848, San Francisco was a sleepy colonial outpost having fewer than a thousand residents. Within a few years, it swelled to over fifty thousand citizens and had become a thriving metropolis of luxury and energy.

The rush to orbit after the development of an inexpensive launch service will be as dramatic as the California gold rush. When people have the opportunity to go into space, they will respond in numbers reminiscent of 1848. Over forty-five thousand teachers requested applications for NASA's teacher-in-space program, and over eleven thousand completed the fifteen-page form. Space Expeditions, Inc., of Seattle, Washington, has been accepting deposits since 1984 for a twelve-hour trip into low-earth orbit, a trip that may not be possible until the next century. Nevertheless, over two hundred vacationers ranging in age from eighteen to seventy have already paid five thousand dollars to reserve tickets costing over fifty-two thousand dollars each. Space interest groups and book sales are another indicator of the depth of the public's interest in space travel. Membership in space interest groups topped three hundred thousand in 1984, the last year for which statistics are available. In California alone, by the late 1980's, the National Space Society had thirteen chapters, published a ten-page monthly newsletter, and had organized a regional council that lobbied Washington, D.C., with a goal "To create a spacefaring civilization which will establish communities beyond Earth." And science fiction has become the largest and most successful genre in American publishing. Twenty years ago, no more than two hundred science-fiction books were published annually; in 1988, there were over seventeen hundred—about 20 percent of mass-market paperbacks. When access to space is as cheap for us as journeys to California were for our ancestors in the mid-1800's, the rush to orbit will eclipse the rush for gold in 1848.

The opportunities for business in the space age will be immense. Today, hundreds of companies consider themselves part of a space industry that hardly exists apart from government contracts. When true markets become possible, the number of space-related businesses will multiply a thousandfold. These businesses will beam solar power down to Earth; they will mine asteroids; and they will manufacture products using alloys impossible to produce on Earth. A large number of enterprises will devote themselves to supplying products and services to space settlers, but the biggest profit-maker may be something no one has considered yet. "I believe," states Peter E. Glaser, a vice-president of the consulting firm Arthur D. Little, in Cambridge, Massachusetts, "that space in the twenty-first century will probably be what aviation, electronics, and computers were, together, in this century."

The rewards for success will be as great as they have been in any other period of our history. Space entrepreneurs will win fame, fortune, and the deep satisfaction of significant achievement. The media will dog these daring men and women to report their historic accomplishments in newspapers, magazines, books, and films around the world. Everyone will know of their deeds. Successful entrepreneurs will earn thousands of small fortunes and a few very large ones. No material pleasure will be beyond their means. And new patriarchs such as John D. Rockefeller and Joseph Kennedy may arise who will use their billions to found families that will influence public life for generations. But the greatest reward for these individuals will be the joy of high achievement. Few pleasures in life can be greater than the euphoria of dreams fulfilled in defiance of limited resources.

The adventures of the young men who built the microcomputer industry in the late 1970's and the early 1980's are a preview of what the future holds for those who can look ahead. Technological advances in microelectronics created several major corporations, hundreds of new fortunes, and a prosperous livelihood for thousands of others. Entrepreneurs earned not only wealth but also fame and glamour. The spotlights of the national media lighted up the microcomputer industry in its heyday as if it were a Broadway stage and made stars of its managers and engineers. As large as the microcomputer industry has become, however, its experience is only a weak prologue to the future. We will see a dozen entrepreneurs like Steve Jobs, a founder of Apple Computer, and another dozen like Bill Gates, the billionaire owner of Microsoft, the software company. These men won fame and fortune by making desktop computing possible for millions. The prizes for those who open space for millions will be far greater and far more numerous.

The solar system contains an abundant supply of the basic ingredients needed to build a new civilization—energy and materials. In contrast to Earth's surface, where the flow of sunlight is regularly interrupted by night and weather, solar power is constant in free space. A solar-power collector within the Earth–Moon space collects on average about ten times the power it would on Earth. Using lightweight, inexpensive mirrors to concentrate the sunlight, more energy can be provided than industry or consumers will ever need. Natural resources are equally plentiful. Because the other planets were formed from the same materials as Earth was, they generally contain the same minerals and metals. According to the National Commission On Space, "In the long run, the abundance of materials in the main asteroid belt is enough to support a civilization many thousands of times larger than Earth's population." The moons of the outer planets contain ten thousand times the material of the asteroids, and the outer planets themselves add another factor of a thousand. A single small asteroid a half-mile in diameter could contain 2 billion tons of high-grade 5-percent-metal nickel steel—a decade's supply for Earth's industry—and more than enough platinum to pay for the entire cost of transportation. The solar system is rich in natural resources.

The greatest uncertainty concerning space settlement is its economic basis. Business cannot exploit the wealth of the solar system until the costs of extraction, manufacture, and transport are low enough in comparison to price to justify risks. Although there are many ideas, no product meets these criteria today except for telecommunications satellites, which are unmanned and can be operated from Earth. Perhaps this situation is not surprising. Before leaving England in 1607 for Virginia, the leaders of the first North American colony compiled a long list of profit-making ideas, none of which worked. Only the unexpected discovery of tobacco assured the economic success of the colony. Later, the Pilgrims in Massachusetts intended to live by fishing and instead became farmers and fur traders. Because economic forecasts are notoriously unreliable, space colonists must be ready to take advantage of local opportunities and be ready to switch when the results are disappointing.

Although the most profitable product in 2000 will probably be something no one has yet considered, space entrepreneurs are likely to find the products and markets to meet their investment goals in manufacturing, mining, energy, remote sensing, or tourism. Communications satellites will continue to be lucrative, and will no longer be entirely controllable from Earth. As orbital locations become saturated with conventional satellites, space construction workers will build orbital antenna farms too big to be launched intact. In 1985, the global-satellite market was worth a total of $4 billion. By 2000, this market may grow to be about $20 billion and support both a launch market of a further $10 billion and a ground-station market estimated at $24 billion.

The environment of space offers unique advantages of temperature, gravity, and vacuum to manufacturing. Simple mirrors and shielding can use sunlight to produce temperatures from absolute zero to many thousands of degrees for smelting, metalworking, and chemical processing. In addition, because vacuum is an excellent thermal insulator, a space factory could be melting steel in one place and liquefying helium a few yards away. The low gravity of space can be used to produce ultrapure products and

products impossible to manufacture on Earth by eliminating the
nuisance of convection and sedimentation and by making contain-
erless processing possible. For example, urokinase is an enzyme
that dissolves blood clots that kill two hundred thousand people
yearly. It now costs twelve hundred dollars per dose. Specialists
believe that quantity production in space will reduce the cost to one
hundred dollars. Finally, so many industrial processes require vacuum chambers at some stage that a significant part of the cost of
many products can be eliminated in the near-perfect vacuum of
space.

Although mining may eventually become the largest space
industry, it may not be feasible for decades. And unless shortages
become acute on Earth, transporting minerals from the Moon or
the asteroids to Earth's surface may never become profitable. However, space mines will be needed to build cities off-planet, because
shipping raw materials from Earth will be much more expensive
than mining it anywhere in the inner solar system. One day space
mining will be as commonplace as it now is on Earth.

> Prospectors and miners will journey out beyond Mars to
> the Asteroid Belt, armed with electronic gear and lasers,
> riding silent electrical rockets rather than braying burros.
> They will be followed by huge factory ships, sailing the
> placid sea of vacuum for years at a time, scooping in
> thousands of millions of tons of rocks and metals, pro-
> cessing them into finished products on the long voyage
> home.

Supplying electrical energy for Earth from satellites intercept-
ing solar energy is the most controversial of possible space industries. Government officials, scientists, and consultants have studied
the concept of solar-power satellites since the early 1970's. The
idea is disputed on both technical and economic grounds. Never-

theless, energy would be an ideal space product with a very large and stable market on Earth, a potential for export sales, and an independence from surface-to-orbit transportation costs, once established. Significantly, the Soviet Union has announced the goal of building the first solar-power satellite to supply energy to Earth in the 1990's. And the largest conference held to date on solar-power satellites was in Japan.

Remote sensing is an existing albeit heavily subsidized industry providing high-resolution photography of Earth's surface. Although it is not yet economically self-supporting, demand for its services grows steadily in fields such as geology, water resources, oceanography, agriculture, and forestry. Some observers, perhaps somewhat extravagantly, consider remote sensing a breakaway technology. A recent publication of the leading aerospace society approvingly quoted two writers who believe that "historians of the future may well compare the development of remote sensing to that of the wheel, so revolutionary and so basic may become its ultimate impact on society."

Tourists will journey into space as costs fall and dangers fade. Enterprising travel agents are already taking deposits for twelve-hour flights in low Earth orbit. By mid-1986, over two hundred travelers had paid five thousand dollars each to reserve seats on passenger flights expected to begin in the 1990's. The $52,200 fee includes registration, a five-day orientation-and-training stay at a resort, the space gear, and the flight itself. A round-the-world cruise on the *Queen Elizabeth II* costs roughly twenty-five dollars per pound of passenger weight, and some aerospace engineers believe that the cost of shuttling into orbit can be reduced to that level by the end of the century. In a typical year, about seventy thousand Americans take luxury cruises; if fifty thousand are willing to visit an "Orbital Hilton" and spend five thousand dollars, the annual market would be $250 million.

But the task of business today does not lie in manufacturing, mining, energy, remote sensing, or tourism. Its task is to drive down launch costs to a fraction of what they are today. Exorbitant transport costs will prevent the development of a self-sustaining, growing human presence off Earth forever. Launch costs now are about what they were in the mid-1960's. Real costs per pound of payload in constant 1964 dollars were twelve hundred dollars for the *Saturn 1* (1965), fourteen hundred dollars for the *Saturn 5* (Moon landings, 1968–72) and fourteen hundred dollars for the space shuttle in the 1980's. These costs are too high. Many worthy projects are impossible because hidden transportation costs make up a large part of their costs. The more it costs to get to space, the less we can afford to go and the less we can afford to do there.

Business must cut the costs of getting into space. But why hasn't it done so already? Why are launch costs still high after thirty years and billions of dollars of expense? Business has not decreased launch costs because it has not had the opportunity. Space transportation is a nationalized industry run by government officials who discourage private efforts. Since the Nixon administration, federal policies have given the government a monopoly in space transportation by setting shuttle prices so low that no private firm can match them without bankruptcy itself; by stalling for years on completing agreements with commercial firms for use of government launch facilities; and by failing to respond to private initiatives that encroached on its domain, such as two attempts to buy and operate shuttle orbiters. The results have been predictable: rising costs, red tape, bureaucratic inefficiency, and little progress in the struggle to make this country a spacefaring nation.

Before business can do its job, before the true power of this country—which consists of the individual energies of our people— can be released, politicians must change the structure of the space program. Our space program is fundamentally flawed. It is not in harmony with our strengths, traditions, or institutions. Despite the good intentions of many civil servants, whose dedication to developing space is incontrovertible, the U.S. space program is mediocre. It has failed us, and it will continue to fail us. It frustrates the energetic, thwarts the daring, and robs us all of our future. If our future in space remains in the hands of the government, the golden meadows beyond Earth will remain the province of a tiny band of supertechnicians employed by governments and megacorporations. If we are ever to realize the hopes space holds for this planet and its people, we must change America's space program. And the men and women who have the responsibility and the power to make this change are politicians.

THE POLITICAL TASK

"Almost all the nations," wrote Alexis de Tocqueville over 150 years ago, "that have ever exercised a powerful influence upon the destinies of the world by conceiving, following up, and executing vast designs—from the Romans to the English—have been governed by aristocratic institutions." Democracies, he explained, are not able "to persevere in a design" or "work out its execution in the presence of serious obstacles" or "await the consequences [of its measures] with patience." If you want to ensure the greatest happiness and the least misery for a nation, he wrote, establish a democracy. But if you want "to constitute a people prepared for those high enterprises that will leave a name forever famous in time, you must *avoid the government of democracy* [emphasis added]."

No experience in American history illustrates the truth of de Tocqueville's words better than the U.S. space program. It suffers from the classic problems a democracy has managing a large, lengthy, and important project. The conquest of space is a vast undertaking spanning decades that has already influenced the world profoundly. The memory of whatever nation is most responsible for leading humanity off this planet will endure for centuries. As de Tocqueville would have expected, the United States has a flawed space program. Policy changes have whipsawed it; problems have hamstrung it; and support for it has oscillated as the years have passed.

The success of the Soviet space program confirms de Tocqueville's point. While democratic America has advanced in bursts of energy followed by periods of inactivity, authoritarian Russia has plodded ahead year after year. The USSR's steady commitment to developing space since the 1950's has made it the world leader. By the late 1980's—despite the fabulous success of *Apollo* and the planetary probes—the USSR had developed the world's most reliable space-transportation system. *Jane's Spaceflight Directory* stated in its June 1986 edition that the Soviet lead in manned spaceflight may be as great as ten years. Russia led the United States in manned spaceflight, launch rate, and heavy lift vehicles. By May 1988, the Soviets had amassed five thousand days in space, compared to the eighteen-hundred days Americans had spent in orbit. The aerospace division of General Motors became so frustrated with the U.S. space-transportation system after the *Challenger* exploded that it asked the State Department for permission to launch communication satellites aboard Soviet rockets.

Some observers believe that the USSR will launch a manned flight to Mars by the mid-1990's. Others think that Russia is poised for a technological quantum jump and a renaissance in the exploration and exploitation of space. If that leap occurs, the United Sates will be further behind the Soviet Union than at any time since the launching of *Sputnik* in 1957. In 1984, the Soviet space work force was already four times that of the United States, and its program was funded at 1.5–2 percent of its GNP, compared to 0.5 percent for the United States. The Russians possess the largest space logistics base and infrastructure in the world, and they also have the world's largest and most active production lines for boosters and satellites. The highly secret Baikonur Cosmodrome is a space complex nine times the size of the Kennedy Space Center. By the mid-eighties, the Soviets had nearly established a permanent presence in space, while the United States space fleet was grounded due to the *Challenger* accident. For years, the USSR has been designing, building, and launching newer, bigger, and better spacecraft. In mid-1987, the Soviets successfully tested the *Energeia,* a 220-foot rocket capable of thrusting payloads into orbit at least four times that of the U.S. space shuttle's orbiter.

In the 1960's, Sergei Korolev, the original leader of the USSR's space program, declared with considerable hyperbole, "The Soviet Union has become the seacoast of the universe." Recalling that the Russians have sought a warm-water port to link their landlocked nation with the rest of the world since Peter the Great, that statement reveals the psychology and strength of resolve behind the Soviet space program. Their progress in the last twenty years and their plans for the future should not fail to impress us.

We ignore at our own peril the commercial and military presence of other nations in a region as strategic as space. Throughout history, the countries that controlled the sea, and more recently the air, dominated international politics. Great Britain's mastery of the world's oceans was the basis of her empire; and the Royal Air Force's successful fight to deny air superiority to the Luftwaffe in World War II frustrated Hitler's plan to invade England. The ability of the U.S. Navy to blockade ports has been crucial to U.S. policy several times in the twentieth century, most notably during the Cuban Missile Crisis in 1962. Space envelops the planet, giving the nation that controls it a commanding position over terrestrial targets and markets. No statesman needs to be reminded of these truths. For the same reason that the young American nation in the eighteenth century did not want British and French warships patrolling its coasts, twenty-first-century America would regret abandoning the high frontier to other powers.

   The force of these arguments is redoubled if space proves to be the economic bonanza that many informed observers expect. After the *Apollo* flights in the early 1970's, the United States had outdistanced all nations in its space technology and could anticipate a similar economic position when commercial development became feasible. That belief proved illusory. The steady progress of the Soviets, the achievements of the Europeans, the new offerings of the Third World, and the probable surge of the Japanese means tough competition. We may lose our leadership in commercial space even before we have it. The United States eclipsed the Old World of Europe through its control of the New World. We have no guarantee that our current prominence among nations will continue in the new worlds of space.

   Scholars cite fifteenth-century China as an example of the dangers of avoiding competition in new regions. In the 1400's, China was the world's largest and richest nation. Decades before Europe sent out its first explorers, large Chinese fleets were exploring the coasts of Asia, Arabia, and Africa. As the United States led the world powers in space after the Moon landings, China once led the world in maritime technology. The Chinese invented the compass and the sternpost rudder; and their ships were much larger and better structured than the comparatively primitive European vessels of the day. But abruptly, their exploration stopped. The Chinese government decided that its funds were better invested in mainland projects such as water conservation for farmers and agrarian financing. When Europe began its great age of exploration, it had no rivals.

   In the 1800's, China suffered the penalty for its shortsighted, insular policies. As Europe grew rich and powerful through its exploration of the New World, China declined. Finally, in the 1890's, China, which had invented printing, the compass, and gunpowder four hundred years before Europe, became the prey of upstart European powers who fell upon her to seize economic privileges and territorial concessions. This ancient and venerable civilization that had a graceful, urban society when Europe was barbaric was looted and divided by nations once too insignificant to secure an audience in its imperial court.

Something is wrong with America's space program. After thirty years of government management, something is very wrong. While foreign rivals have moved steadily forward, the government has mismanaged the manned space effort; it has alienated its partners in business and science; and it has allowed the military to take much more than its share of scarce budget dollars. Although we can find several bright moments in the last three decades, these individual achievements should not confuse our view of the whole record. We should not judge American performance in space by the few gains purchased at the cost of tens of billions of dollars, but by how much more we could have accomplished, how many opportunities we have discarded, and the sorry condition in which we now find ourselves.

The greatest loss has been in the manned space program. The government's man-in-space projects run from the *Apollo* Moon landings in the 1960's through the shuttle in the 1980's and the space station planned for the 1990's. The record is a poor one. *Apollo* was an empty triumph that left us with nothing in space and little on Earth. The space shuttle was delivered far later, costs much more, and flies far less than promised. The costs of the space station are already ballooning, and its deadlines lengthening. We can measure our loss by how far the USSR—a nation widely recognized as our inferior in developing new technology—has outdistanced us. Had the Moon program given us an enduring infrastructure, had the Nixon administration not gutted the shuttle plans, and had we continued to send probes to explore the solar system, humanity would be much further along its path today.

Federal policies have outraged businessmen and scientists. Business, which needs a stable investment environment, has suffered from eight shifts in aeronautical-communications policy, five shifts in satellite-communications policy, eleven shifts in remote-sensing policy, and five shifts in satellite launching-service policies. In addition, it must answer to thirteen federal regulatory bodies, two international organizations, and four treaties. Given this background, any commercial interest in space is surprising. Space science is in a state of crisis because the government has concentrated on huge projects such as *Apollo* and the shuttle. The relatively small investments required to build upon the wonderful planetary probes of the 1970's were shoved aside by the enormous cost overruns of the shuttle. The government broke commitment after commitment, delaying some projects over ten years. Continuing problems are driving veteran scientists elsewhere and discouraging graduate students from entering the field.

But the greatest shame to American citizens is not the loss of our leadership in space to the USSR or the steady approach to parity of Europe and Japan. Such rivalry merely wounds our pride. Our greatest shame is the domineering presence of the military in our space program. For a people who could once boast of a space effort untainted by military influence, who could once see ourselves as envoys for all humanity, the militarization of space is shameful. The Soviet Union has had a secretive space program from the beginning, while the United States has conducted its own program under the scrutiny of international publicity. No longer. Since U.S. military spending in space jumped 85 percent between 1980 and 1987, more and more launches bar the press for national-security reasons. Some two thirds of the funds Congress authorizes for space now go to the military.

Why is the U.S. space program failing? It is failing because it is based on a mistaken understanding of what this country can do. We are not a people with a history of accomplishing our purposes through government. In fact, we are part of a tradition that distrusts government and its powers. We are descendants of immigrants who fled the restrictions of their native lands to become citizens of a country too big and too diverse to be ruled by central authority. During most of our history, the national government has been weak compared to other sectors of American society, because the Constitution created institutions that check each other's power. Not until the Depression and World War II threatened our national existence in the 1930's and the 1940's did the federal government win its current preeminent position. But we still have not abandoned our bias against government. Most of the presidents since World War II have been members of a party that regards reducing the role of government as a central part of its mission. We dislike big government.

This antigovernment tradition means that we are not a nation with much experience in building institutions that can pursue ambitious public objectives for long periods of time. Of course, the federal government has been responsible for many beneficial measures in the last two hundred years. But the primary field of our activity has never been governmental. We can act through government—as we have demonstrated on numerous occasions—but we are not particularly good at it. And if we believe de Tocqueville, we are less capable of public action compared to nondemocratic regimes. We do not have the stability. We are an energetic, mobile, changing, and self-centered people; and our institutions reflect this character.

The government itself recognizes its limitations. The Office of Technology Assessment is a respected staff arm of the House Committee on Science and Technology. In a recent technical memorandum, it reached the following conclusions:

> . . . the institutional structure and will [of the government] to focus the efforts of [the] interested parties on the common purpose of reducing operations costs does not presently exist. . . . The current institutional struc-

ture . . . is poorly structured to lower launch costs or increase launch rate. . . . *routine, lower-cost access to space . . . is probably unattainable unless the U.S. Government substantially alters the way it conducts space transportation operations. . . .* [emphasis added]

What, then, is the alternative? On what other principle can we base the space program? We can base it on an ability that distinguishes us from other nations, an ability we have in excess that some nations do not have at all. We have this gift: We know how to liberate the full powers of individual talents for common goals. We know how to release the energies of individuals and harness them to social purposes. It is this ability that permitted us to build a nation out of a wilderness and an economy richer than any in history. It is this ability that is the source of our world-renowned spirit of entrepreneurship and innovation. We have this capacity because we are a nation of individuals who seek self-fulfillment in defiance of authority, convention, and tradition. We are not Russians used to obeying rulers, or Japanese obedient to custom, or Europeans weighed down by a venerable heritage. We are Americans—living in a land that teaches us early to be independent of society, family, and past, and believing that each child starts anew in the quest for success and freedom.

The political task is to restructure the space program so that it is in accord with the fundamental political processes of this country. We are a nation of individuals with dreams of economic success and personal freedom. The politician's job is to harness these powerful drives in service to our need to make space accessible to everyone. Let other nations develop space through their governments. We are mediocre at creating giant institutions to accomplish distant purposes. If we continue to try, we will not be using our strengths, and nations that do know how will soon surpass us. We should do what we do well: unleash the energies of millions of individuals to explore, settle, and develop new frontiers. Only then will we prosper as a nation and contribute to the advancement of our race.

HEROISM

Not too many thinkers in our time reflect upon the nature of heroism. We live in a commercial era in which economic man has replaced the citizen of civic virtue. We dream more of the leisured enjoyment of wealth than of the exacting discipline of public life pursuing the common good. Few want to be noble, and fewer know what it is.

One who has considered what it means to be a hero is the California historian James S. Holliday. Mr. Holliday believes that we have lost much by forgetting our national heroes. He has compiled his own list of American figures "whose character sets an example of how life should be lived," men and women who knew how to stand alone on principle. Included in his list are traditional figures such as George Washington, Thomas Jefferson, Benjamin Franklin, and Abraham Lincoln; modern leaders such as Louis Brandeis, Fiorello La Guardia, and Martin Luther King, Jr.; and writers such as Bernard De Voto, William Allen White, and Rachel Carson. These people led lives that, in Holliday's opinion, are "a response to a sense of duty, of service, of striving for the common good." They are heroes with "that kind of character that will sustain over a long period of time the courage to pursue a goal, to champion a cause in the face of the dangers of ridicule, vindictiveness, even ostracism."

The heroes of our generation will be those who help open space to the peoples of the world. No other act can contribute more to the common good. What other action offers us more hope of diffusing the crises threatening the planet? What other action gives us more promise of preserving American traditions amid fierce international rivalries? What other action gives us a better chance of finishing our lives in economic security and passing on to our children a quality of life at least equal to our own? To serve the cause of space settlement is to serve humanity, our nation, and ourselves. If the heroic means anything, it must refer to acts with consequences as beneficial as those that will follow from opening the space frontier.

There are many ways to participate in this enterprise, and at different times different roles will be important. Sometime soon we will need brave families to pioneer new communities. The wilderness they will encounter will have hardships, dangers, and joys that are unknown to late twentieth-century middle-class life. At some other time, we will need explorers willing to endure privations for years in journeys to worlds far beyond frontier settlements. Among their rewards will be the discovery of new natural wonders that will rival those of America's first frontier, such as Yosemite Valley and the Grand Canyon. At other times, other occupations will become significant. Making humanity spacefaring requires the contributions of many.

At this moment in history, one task stands before all others in importance—the development of a mass-space-transportation system. The settlement of space will remain an uncertain project until this deed is done. Before explorers can start their adventures and before settlers begin their communities, space must become more than an extreme and desolate verge of human life. It must become part of the human domain. It must become part of our ordinary lives. When trips to space become as cheap for us as journeys to the New World were for seventeenth-century Europeans, *Homo sapiens* will have entered a new age. Then the departure of millions from their home planet to start new lives throughout the solar system will unleash forces that will counteract the nemeses that threaten us here.

The agents of this momentous change are engineers, businessmen, and politicians. The practitioners of these professions live now in a golden age, because from their ranks will come the men and women who will be the heroes of our age. Rarely will the professionals in any field have the opportunity of doing so much for the rest of us. Only the politician can form the national will to restructure this country's space program. Only the businessman can create the enterprises that will provide cheap and reliable mass transportation to the heavens. And only the engineer can devise technologies that will reduce launch costs ten thousandfold. All our hopes rest with these professions and with those who practice them. They will make humanity a spacefaring species.

"Perhaps no form of government," wrote James Bryce in his classic study *The American Commonwealth,* "needs great leaders as much as democracy." We need leaders now. The planet suffers and new worlds beckon. Surrounding us are dismaying problems that chill our hearts, while above us is a future that is wonderful. But the opinions of some ("You really have a hard time making a case that there is anything unique about the [postwar] generation."—Warren Miller, professor of political science at Arizona State University) may one day become the judgment of history. We need leaders—men and women committed to high public service—who can awaken the civic activism that flourished in our youth, create the political will to restructure our nation's space program, and stimulate the development of cheap access to space. Our task is clear. We have a new movement to form, laws to change, companies to build, and technology to develop. We must make space open and free. That is our opportunity; that is our challenge. "Certainly all historical experience confirms the truth," wrote the sociologist Max Weber, "that man would not have attained the possible unless time and again he had reached for the impossible." Heroism is possible; it is desirable; it is necessary. This is our moment. Let us make the most of it.

CHAPTER SIX

The Role of the
Gifted Generation

We of the gifted generation can make this moment in history ours by beginning the settlement of the solar system. The public philosophies of the right and the left are losing their appeal, and no other goal offers us as much hope as the mission to open space for human habitation. An expanding human presence off this planet will alleviate Earth's crises, strengthen the United States, and make all of us more prosperous. America has centuries of frontier experience and is the favored offspring of the most expansionary civilization in world history. We are a nation of explorers, settlers, and entrepreneurs. And we of the postwar generation—the direct heirs of this magnificent tradition—have the power to move our country forward and to grasp the grand chance before us. We have the numbers, time, and skills to make the 1990's the start of an era that surpasses what we accomplished in the 1960's.

The opportunity we now have may be fleeting. We face crises that can make this planet unlivable. But if we can survive and become spacefaring, our future is grand. We are a species blessed with talents that have given us control of Earth in an extraordinarily short period of time. We live on a planet bursting with life in a vast galaxy that appears devoid of any other intelligent creatures. No other spacefaring species may exist within a million light years. Fortunately, the record of innovation in the last two centuries suggests that the technology to move human life into space will be available soon. When it is, we can begin the task of building a solar-systemwide civilization in which a regenerate Earth is the crown jewel and this solar system the seedbed of life for the rest of the galaxy.

THE POWER OF THE GIFTED GENERATION

Our gifted generation is responsible for whether the twenty-first century becomes a threshold to a broader existence for humanity or an era of ruinous catastrophes. Beginning with the 1990's, we will dominate American politics for decades. In the same way that we defined the teen culture in the fifties, the student revolts of the sixties, the social movements of the seventies, and the consumerism of the eighties, we now have the power to set the political agenda of this country. We are now leaders of Earth's greatest superpower and have at our disposal all its resources. How well we use them may determine the fate of this planet.

This view of the role we can play in history is not an exaggeration. No one can foresee how nuclear war, overpopulation, and environmental disaster will affect us in the next hundred years. The effects could be catastrophic. On the other hand, the presence of such unprecedented dangers may stimulate us to improve our dismal record of intraspecies cooperation in order to survive. One consequence is certain: We will lose the opportunity to become spacefaring and to people the galaxy. Making humanity extraterrestrial involves enormous costs, because major extensions of life are always expensive. Huge sums must be spent for research, development, and construction; and vast resources must be redirected from other uses where the return on the investment will not be as distant and uncertain as it is for space ventures. In one way or another, each of the major problems facing us tends to reduce the resources available for space and to make our political life more complex. We cannot look skyward if our attention is absorbed by our troubles on the ground. A world with a population greater by a few more billion and with a few thousand more sophisticated nuclear weapons; a world beset by flooded coastal cities or radiation from meltdowns at nuclear plants; a world faced with an escalating scarcity of coal and oil—such a world will not look favorably upon journeys into outer space. If humanity is to realize its promise of becoming the life-form to settle the Milky Way, we must act now. We may have no other chance to avoid being chained to this planet forever.

The number of people in our generation provides us with the primary source of our power to move the nation. Between 1946 and 1964, over 76 million people were born, about one third of the total U.S. population in 1980. Because our generation forms such a disproportionate share of the population, we have influenced society disproportionately. Whatever we wanted, we got. As infants, we caused diaper sales to shoot up by 50 percent. As children, we stimulated the toy industry to produce us new playthings, such as Silly Putty, Davy Crockett caps, Slinkys, Hula-Hoops, and Barbie dolls. As youths, we demanded blue jeans and backpacks. We also forced society to pay attention to our concerns: We defined each era, from sex and politics in the sixties to parenting and housing prices in the eighties. Our numbers have always compelled attention.

Our influence over the economy and society in the first half of our lives foreshadows the sway we will have over national politics in our maturity. In the 1990's, we will head every other household in America and form a majority of the electorate. We will be responsible for over 55 percent of consumer spending and over 54 percent of the labor force. We can do much with these resources. A look at the American Association of Retired Persons (AARP) in 1988 gives us a glimpse of how powerful we may become at the ballot box. The AARP had 28 million members that year—more than most nations. In the 1988 New Hampshire presidential primary, it mailed 250,000 pieces of literature in a state in which only 101,000 voted in the 1984 primary. Its New Hampshire membership alone totaled 145,000. Consider how the influence of organizations like this one will grow as the percentage of the population above sixty-five climbs from 12 percent in 1988 to 21 percent in 2030. Money and votes decide political issues in this country, and we will soon control both.

We have received one final gift from our era: longevity. This gift dramatically increases our ability to change the career of this country. We will be the longest-lived and healthiest generation in history. Life expectancy has been rising steadily and may increase even faster in the future. When the Constitution was written, Americans could expect to live only about thirty-five years. But by the last half of the nineteenth century, life expectancy had reached sixty years; and by 1980, it had climbed to over seventy-three years. Seventy percent of the men and 80 percent of the women of the postwar generation will live to be sixty-five. Demographers also believe that those who do attain that age may live fifteen to twenty years longer. According to some experts, one third of us will live to be eighty-five or older. In addition, we are living healthier lives. Fully half of those today between seventy-five and eighty-four are free of health problems that require special care or that curb activities. At least one third of those over eighty-five suffer no limitation due to health. And these percentages will certainly rise in coming years as age-related research progresses. The nonprofit Alliance for Aging Research surveyed fifty top scientists in 1987 and reported that "within the next 20 years the means may be at hand to avert or postpone major diseases of aging"—such as heart disease, osteoporosis, and Alzheimer's disease. By 2000, the number of people eighty-five and above will triple. Not only do we form a disproportionate share of the population in the closing years of the twentieth century, but we will also retain this position for at least the first decades of the twenty-first century.

Our long lives give us enough time in which to make a contribution to the world commensurate with our gifts. We may have wasted precious years, and we may have made foolish mistakes— but we still have time. The senior members of our generation will be in their forties in the early 1990's and will still have decades of active live before then. No one need fear that older citizens cannot contribute to the commonwealth. In fact, the political strength of the old exceeds the influence of their numbers and a voting rate higher than any other segment of the population, because they are mature people who have more time for public issues than younger citizens. Their children are grown and have left home, and they have usually achieved some measure of financial security. They even need less sleep, since men and women over sixty-five often sleep only three to six hours a night. Just as we created a youth culture in the 1950's and 1960's, we will transform the America of the latter half of our lives into a nation in which age is respected and admired.

We may one day look back upon the 1960's as the time when our political ability first became manifest. In those years, we exhibited an appetite for politics that was absent from previous generations of Americans. They had other challenges. They fought wars or survived depressions or settled frontiers. The task we faced was to change public policies that we believed were wrong. We knew what we believed in, and we had the courage to support our convictions by doing something. Because we were committed to racial equality, we marched in the South for civil rights. Because we were committed to peace, we marched on our campuses and in the streets against the Vietnam War. Because we wanted to serve, thousands of us joined VISTA and the Peace Corps. In the 1970's, other issues affected us. Some of us started the environmental movement to stop industrial practices harmful to global ecology; and women mobilized for their fair share in the economy and in the home. The early history of our generation is a history of devotion to a few high causes that compelled the nation's attention and led to changes in public policy. "We ended a war, toppled two Presidents, desegregated the South, [and] broke other barriers of discrimination," boasted Tom Hayden in 1977. We recognized the necessities of the times and did what we could to shape a history of which we could be proud. The 1960's were a striking demonstration of this generation's remarkable political abilities and will perhaps be recognized one day as only the first phase of our political activity.

The cry for a revolution that millions of us uttered in the 1960's and *The Port Huron Statement* that the newly formed Students for a Democratic Society issued in 1962 illustrate the heights to which our idealism could climb. Neither deserved the reputation for radicalism that it later acquired. The revolution we yearned for was never the bloody overthrow that took place in Russia and in France. We never wanted to replace the Constitution or discard the Declaration. When we spoke of revolution, we were evoking hopes of beneficial change for the nation and only wanted to make the future much better than the past. We only wanted to become renowned for uncommon achievements in service to the common good. *The Port Huron Statement* expresses equally ambitious sentiments. Its authors intended to describe an "Agenda for a Generation" in which was set forth a vision containing "the perimeter of human possibility in this epoch." Although the activists who wrote these words may have fallen short of these goals, they did succeed in underlining many of the important issues of the following ten years, such as civil rights and peace. And they did so within a context of traditional, mainstream Americanism. For by its passionate interest in democracy, individuals, and purposeful action, *The Port Huron Statement* marks itself as a classic expression of American patriotism.

We were willing to back our beliefs with deeds. The force of our convictions for change moved us beyond ideals to action. In the 1960's and 1970's, millions of us joined mass movements and service organizations that fought for social change. Out activities were not cost-free. They required commitment, and they sometimes demanded courage. Tom Hayden first encountered a mob when he was working in Fayette County, Tennessee, in early 1961, for civil rights. "We were walking down the street and we ran into a gang that was just waiting for us with belts and clubs. We walked the other way as quickly as possible . . . and called the cops. But they just joined the toughs." Later that year in Mississippi, he was dragged from a car and beaten. His experiences were not unique. In the antiwar years that followed, the police clubbed demonstrators with nightsticks and rammed them with rifle butts; they smashed heads, limbs, and crotches; and they used tear gas and shotguns. Arrest and jail were the least of our worries.

The experiences we had struggling together for important issues will affect us for the rest of our lives. In contrast to the apathy characteristic of modern democracies, we experienced in the 1960's the intensity and excitement that can come from selfless devotion to a great cause. We did not feel that we were powerless or that our actions were meaningless. We were contributors; we were participants. When we marched in a demonstration or when we occupied a university building, we were braving official force and hastening the end to an unjust war. The political activity of our youth gave us an understanding of the possibilities politics may have for us as individuals. "A mass Movement to change America briefly flourished," wrote James Miller, author of a history of the 1960's, and "the sense of what politics can mean will never be quite the same again."

The opportunity to reshape the world again lies in our hands. We are a generation with a collective social conscience, a collective sense that we can do great things. In no other nation does the ruling class have the numbers, gifts, or resources we have to build a hopeful future. And we are mounting the national stage at the same time that the ideologies that have guided the United States for the last half-century are collapsing. We have the power to lead our species off this teeming planet, out into an awaiting galaxy, to begin the construction of an extraterrestrial civilization. The opening of a space frontier is the most important issue of our times. It addresses our most pressing private and public concerns, because it offers us our best chance to increase our personal fortunes, preserve the nation, and avoid global catastrophe. This generation has a great opportunity to live up to its promise by helping humanity evade the multiple dangers threatening it, at the same time that we are strengthening our country and enriching ourselves. The 1990's will renew and upgrade the 1960's. That earlier decade was an immature exercise of our embryonic power and idealism; the nineties will be far different. The revolution is still possible.

THE MOST IMPORTANT ISSUE OF OUR TIMES

No issue has a stronger claim on our consciences and our energies today than the effort to make humanity a spacefaring species. Exploring the solar system and building an extraterrestrial civilization is a moderate, sensible response to the dangers and opportunities we face. Not only are we nearly trapped by the triple crises of overpopulation, ecological pathology, and nuclear war, but we also stand on the verge of an explosion of life throughout this solar system and into the galaxy beyond. Should we escape the catastrophes threatening our home planet and succeed in creating a multiworld civilization, we shall possess a human habitat in which our enormous fertility is a boon, in which Earth will regain her pristine beauty, and in which militarists can no longer menace life with total destruction. But we must act now. Time is running out.

Perhaps some combination of disarmament, nonproliferation treaties, embargoes of crucial materials, and preemptive strikes against nuclear facilities in Third World countries will preserve the nuclear peace forever. The nations of the world should use every resource they have to avoid nuclear war, and the construction of a civilization that spans the solar system must be part of this effort. In such a civilization, nuclear war will still be a danger, but political and military leaders will not have the power they now have to destroy all life. Humanity will have moved back from the abyss of self-destruction. Should nuclear war break out on one planet of a multiworld civilization, life could continue uncontaminated on the others. Moreover, the aggressions that are now cooped up on Earth and intensified by the limitations of a closed planet will be diluted. There will be alternatives to fighting, and fewer reasons for it.

Most people in the nineteenth century would have regarded as ridiculous a prediction that hundreds of millions of passengers would be transported through the air in the twentieth century. Since such a forecast would have become true, we should not be skeptical of forecasts predicting hundreds of millions emigrating from Earth in the twenty-first century. We are a people who progressed from the Wright brothers' fragile biplane to mass air transportation in less than sixty years. Earth today, with her desperate billions, resembles Europe with her desperate millions before she sent emigrants around the world. The explosion of life into space is coming soon. After beginning in the seas, life moved to every piece of land on Earth, from the arctic to the tropics, and it is now ready to move to other worlds. The project in which we are engaged is as ancient as life itself. It will not stop now.

According to the Gaia hypothesis, the earth is a single organism, just as much alive and whole as the smallest unicellular animal. Since the British scientist Dr. James Lovelock formulated this theory in the early 1970's, more and more natural scientists have recognized its validity. In this view, the loss of topsoil, enlargement of deserts, depletion of fishing stocks, increase in acid rain, and accumulation of atmospheric carbon dioxide are diseases. The earth is sick, sick with illnesses that cannot be relieved by ordinary cures. The path to health is not through irregularly applied, partial remedies, but through a fundamental reorientation that makes Earth an exporter of life into space. Only when life again radiates outward and has stopped its insane ingrowth on this closed planet will Earth be well again.

The benefits of life's expansion into space do not stop at survival. We also have much to gain as a nation. Our country is by nature and history a frontier society. We are explorers, settlers, and entrepreneurs. We have an economy, a government, and a foreign policy derived from our experience of more than three hundred years as a nation with an open frontier. Sometime in the last half-century, that frontier ceased to exist. Without it, we no longer have the constant outpouring of new wealth invigorating the economy; without it, we no longer have the continual creation of new men and new institutions democratizing our public life; and without it, we no longer have a counterweight to our hereditary clumsiness in conducting international affairs. If we open the space frontier as the West was once open, our best years as a nation lie ahead. If we do not, America will not survive in its present form.

The solar system contains more opportunities for winning wealth and maintaining our living standards than have been seen since the Industrial Revolution. The lands in space are not barren, inhospitable Antarcticas on which only camps of scientists can dwell. On the contrary, we most fittingly compare the fresh worlds of the solar system to the New World of America that awaited seventeenth-century colonists. New England in the 1600's had its own formidable hardships and its own terrors, yet successive generations of immigrants transformed it into a region enjoying one of the highest living standards ever. The wealth in minerals and energy available for our use in space makes small any terrestrial comparison. Our solar system contains the resources to make a human life possible for far more people.

No issue has greater moral leverage than space. Transforming

our species into a spacefaring creature may satisfy more ambitions
than any other pursuit, whether we want to work for humanity's
common concerns, serve our country, or win wealth and renown.
Expanding the human domain beyond this planet çan help preserve
the species, strengthen the nation, and make our individual lives
more comfortable. In contrast to other great moral issues of this
era—ecology, human rights, and arms control, for example—
opening the space frontier may have wide-ranging and beneficial
consequences far beyond its immediate effects. The exploration
and settlement of the solar system is a project that addresses the
most pressing private and public concerns of our times. Because
building a solar-systemwide civilization has many possibilities
wrapped up in it, no issue has a stronger title to our consciences.

THE MISSION OF THE GIFTED GENERATION

One of the most remarkable features of our predicament is the
extraordinary conjunction of circumstances that may permit our
race to do much more than survive. It is astonishing that at the
time of our species' greatest danger, a technology, a people, and
a generation exist that can triumph over our adversities. These
resources need not have been contemporaneous. First, space may
never have been a practical alternative. There is no reason why
we should have developed the technology to spread life beyond
this planet at the very moment when global conditions demand
that we do just that. A technological civilization such as ours is
not necessarily spacefaring. Second, the world's leading nation
could have been a traditional superpower. There is no reason
why at this critical moment the United States should be a nation
that is the chief heir of the most expansionary civilization in history, excelling in the development of frontiers. In any other era,
the world's leader has been more skilled in conquest and dom-

ination than in exploration and settlement. Finally, the generation rising to power in this country could have had less cause to devote itself to high achievement. We of the postwar generation, however, have the knowledge of our gifts and failures to spur us forward. We know that we have been given much and have much to return.

This country's record of innovation in the last two hundred years suggests that technology to make mankind spacefaring may be near. "Americans," state two highly regarded historians, Henry Steele Commager and Allan Nevins, have "probably patented more numerous and more ingenious inventions than any other people." Between 1900 and 1970, more advances occurred in technology than in all history; never has there been such an outburst of invention. The record of innovation in aerospace alone is significant. The first pilot to break one hundred miles per hour was a Frenchman named Jules Vedrines, who set the world's record at 100.22 mph in 1912 in Pau, France. The *Apollo* astronauts needed a velocity of about 25,000 mph for their flight to the Moon in 1968, and the *Voyager* spacecraft exploring the outer planets travel at over 40,000 mph. This record of achievement does not guarantee the development of mass space travel, but it does indicate that the forces of innovation in our culture are powerful. Technology will not stand in the way of colonizing space, says the eminent scientist Freeman Dyson.

America is the principal heir of a civilization that has expanded more rapidly and more widely than any in history. In the fifteenth century, the West was a backward set of squabbling feudatories occupying an outcropping of Asia called Europe. At that time, proud civilizations flourished in both the Old World and in the Americas. Five hundred years later, Western civilization dominates the planet with its culture, politics, and military, while its onetime contemporaries and former rivals have withered. The United States is the favored child of this grand civilization. We have derived almost everything of what we are from European sources, and our own history has mimicked that of our parent. We moved across the continent from the sixteenth to the nineteenth centuries, and in the twentieth century we landed six expeditions on the Moon and sent probe after probe into space to survey other planets. No nation has ever had the skills we have in developing new frontiers, because none can match either our heritage or our training. We are a nation of explorers, settlers, and entrepreneurs.

National policies leading to a space frontier accessible to ordinary men and women satisfy the deepest needs of the generation that will assume leadership of the United States in the 1990's. We of this generation—the postwar generation, the gifted generation—have much to do. We must live up to our own expectations of what we believe we deserve. We must build an economic life that will provide sufficient opportunity for all Americans to maintain a general prosperity. We must build a political life that suits our sense of social justice and that addresses the grave realities facing life on Earth today. And we must find some way to compensate the country for the harm we caused by popularizing cheap drugs and free sex. If we can create a frontier above us that is as much a part of American life as the western frontier once was, we will have gone far toward meeting these responsibilities and fulfilling the hopes that we once raised.

The historian Arthur Schlesinger, Jr., believes that American history has thirty-year cycles of liberalism and conservatism. The 1890's, 1920's, 1950's, and 1980's were periods of private purpose, when Americans devoted themselves to making and spending money. During these times, we admired the plutocrats who amassed wealth and the celebrities who knew how to spend it. Years of public purpose followed the first three periods, and there is reason to believe that the 1990's will also be a time when common goals move the nation. In the 1990's, our generation will have reached the prime of life—this generation that has left its mark on each era through which it has passed. Some of us will be advancing toward the peak of our professions, while others will begin to occupy positions of leadership in society. Two decades of private life will have matured us, deepening our·understanding of ourselves and our world. We will be ready to become the leaders of the country. In the nineties, we will renew the spirit of the sixties, when our idealism and power first stirred the land. Our time will have come.

The task of orienting America toward space and making our species spacefaring is the historic mission of this generation. We have the numbers, time, and skills to make it happen, as perhaps none of our descendants ever will. And we are emerging as a political force at the same time that the public philosophies that have dominated American politics for a half-century are unraveling. Humanity—young, vigorous, fertile, and inventive—is bursting the bonds of this tiny planet that becomes sicker by the year, while an empty galaxy awaits the magic of our life. In the same way that we shaped the 1960's, we of this gifted generation can provide the leadership and momentum the space movement needs in the 1990's. Space is the issue of our maturity, as civil rights and the Vietnam War were the issues of our youth. It is the most important issue of our time. It is the great opportunity that stands before us to grasp.

CHAPTER SEVEN

Breakout
into Space

We must get off this planet.

The time has come to recognize that the planet Earth is no longer large enough to contain humanity. We stand heartbeats away from extinction through nuclear war. Our numbers increase geometrically, while land and food supplies diminish. Ecological problems are everywhere: air pollution, acid rain, topsoil erosion, water shortages. Worse still, large-scale ecological disasters—such as Love Canal, Three Mile Island, Bhopal, and Chernobyl—have become more common. Finally, the great democracy of the United States, the refuge of the world, falters. Its economy slides, as families disintegrate and drug abuse grows among its citizens. By nature and history a country of explorers, settlers, and entrepreneurs, America is inept in its unaccustomed role of imperial world power. Without a frontier, its democracy is no different from every other short-lived democracy in history. It cannot endure.

The political crisis that faces this generation is despair. When we look at the world around us and the future we face, we see little cause for optimism. Every generation considers itself special, but this generation lives in circumstances that seem uniquely unfortunate. We resemble the condemned on death row in a penitentiary, not knowing when our appeals will be exhausted. The executioner may be an exchange of missiles by the superpowers, nuclear terrorism, an unforeseen ecological disaster, or a gradual loss of our prosperity and liberty in a world packed with hostile masses. On death row, the luckiest of the condemned may be those who live out their natural lives before the bureaucracy of justice has reached a final verdict. If we remain an earthbound species, that fate may be our best hope too. We cannot escape from our own death row to the comparative innocence of life before the modern age. We can only break out into space.

The solar system is ours. No other form of life exists to claim the bounties of a civilization that spans the solar system and eclipses Earth's present troubles. The wealth available in space can provide lives of abundance for a population several times that of Earth now. Nor can normal human belligerence extinguish a species dispersed throughout the solar plain, enjoying lives much more humane than those lived in late twentieth-century megacities. And we will regenerate the ecology of our home planet when humanity's home is the entire solar system. The air will be cleansed, the rains purified, and the extinctions stopped after we have enlarged the human habitat to its natural domain extending from Mercury to Pluto. Earth will become a garden again.

Our breakout into space is not an abandonment of this planet. Centuries will pass before Earth ceases to be the vital source of life for the galaxy of stars in which we dwell. Without Earth, colonies in space will wither as the fruit on a tree whose trunk has died. But to protect the mother planet from the apocalyptic forces that bedevil it, we must build an interplanetary civilization. The problems of nuclear war, overpopulation, environmental damage, and political oppression will never be solved by a humanity pent up on this small rock. We must begin the movement of Earth's life into the rest of the universe. The enlargement of territory, wealth, and spirit that all nations of the world will experience in space is far more likely to preserve the species than the restrictions of a planetbound mankind sweating on increasingly barren, increasingly smaller patches of ground.

We live in one of humanity's great ages. Our powers, our perils, and our possibilities are at a peak, matched rarely in the past and perhaps never again in the future. Our descendants will remember the twentieth century as an era when fateful decisions were made that had consequences for centuries. Deadly crises besiege the planet at the very time that life outside of Earth's atmosphere has become possible. We live in a galaxy that has countless new homes for life. And it beckons us as a new outlet for mankind's energies, offering living space, economic self-sufficiency, and the hope of decent human lives for trillions. But the next fifty years are critical, because the establishment of our permanent presence in space must take place in this period. If it does not, the instability of the global political structure and the steady diminution of Earth's resources may make the task impossible.

We have a grand opportunity. After eons of evolution, a life-form that can green the stars has emerged on a small planet orbiting an obscure sun in a galaxy largely hostile to life. Despite enormous odds, Earth teems with members of the species *Homo sapiens sapiens,* while the probability of intelligent life existing elsewhere on any of the galaxy's large number of habitable planets is negligibly low. Humanity is alone in the Milky Way, but it is alone with the ability to spread life throughout the galaxy. The technology now exists to create orbiting colonies the size of small European countries, with sunlit meadows, trees, and lakes, housing communities with millions of people. Using the sun's energy and minerals from asteroids, these colonies can become self-sufficient arks capable of traversing the solar system. The fires on Earth are burning out of control. A vacant solar system lies at our doorstep, and a thousand new suns beyond the horizon. Let us pour forth into space, build an interplanetary civilization, and make Earth an Eden once more.

The survival of Earth rests with our country. Among all the

nations of the world, only the United States has the skills and the resources to enlarge the human habitat before the bombs fall. Were a new Homer to imagine a people suited to the task before us, he would conceive a nation such as America is today. "This is the American moment in world history," declares Allan Bloom in *The Closing of the American Mind,* "the one for which we shall forever be judged." America is the heir of the best traditions of Western civilization, the same relatively backward culture that was confined to Europe, a small peninsula of Asia, as late as the fifteenth century. By nature imperial and colonial, the West broke out of Europe, explored the world, and grew to dominate it. As a people with over three hundred years of frontier experience, America is its true son. America possesses the technologies and the traditions to move humanity off this planet, as the English and the Spanish once moved Western civilization out of Europe. But America is not just European; it is African and Asian as well. Immigrants from every land have made the American nation the representative of all humanity. America *is* humanity. When Americans first land on Mars, when Americans first pass beyond the asteroid belt bound for Jupiter, when Americans travel out of the solar system on the first starship, the triumph will not be narrowly nationalistic. It will be a victory for all mankind.

But America needs a new public philosophy—one that captures what we are and what we can do as a nation. Ideologies conceived a half-century ago in the Depression cannot guide us in a postmodern world having problems and possibilities unknown in our parents' youth. Late twentieth-century America should know now that it is not an imperial nation fit for global rule but a people formed by more than three centuries of frontier life. We are the offspring of the most expansionary civilization in history, and our own history reflects that heritage. We explored, settled, and developed a continent, and we have now launched interplanetary probes that continue a tradition of exploration that began before

Columbus in the fifteenth century. We can never act like Roman proconsuls, British colonial governors, or Russian commissars. We need a public philosophy that recognizes our weaknesses as an international power and builds on our strengths in innovation and individualism. Such a philosophy will promote large-scale extension of life beyond this planet through the establishment of self-sufficient settlements in which human life and democratic traditions can flourish. "[Dr. Gerard] O'Neill and I have a dream," says Freeman Dyson, "that one day there will be a free expansion of small groups of private citizens all over the solar system and beyond."

No other group is as well positioned as the generation born between 1946 and 1964 to lead humanity off this planet and into its new realm. Despite our failures in the economy and private life, we have a promise that will one day shine forth. We are the gifted generation, raised in affluence, recipients of a lengthy education, and capable of high political achievement. The grand opportunity before all humanity is our grand opportunity, and few are placed as well as we to benefit from it. And we have many years in which to do our work. Let us make the 1990's the beginning both of our leadership of the country and of a new era for the nation. Settling space satisfies our deepest needs. The task of this generation is to lead America into space.

We live now on the threshold of a heroic age. After a time that idolizes entertainers, after a time when great souls appear to have vanished, mankind is again entering an era when great deeds are possible. Wonderful discoveries will be made, dangerous voyages attempted, and impossible perils overcome. Wealth that will found dynasties and fame that will endure for centuries will be won. We stand now in that slow transition between epochs when old ideas and ambitions are fading and being replaced by new ones. Each year our confidence in humanity's new direction grows stronger. Each year we see more clearly that the time when humanity must leave this planet to begin its migration through the Milky Way has arrived.

The purpose of this book is to persuade the choice and adventurous spirits of our age that the noblest enterprise and the highest purpose in our time lie in efforts devoted to the settlement of space. No matter how profitable, laudable, or humanitarian other movements may be, none can compare with the moral grandeur of the possibilities unfolding before us in the next few decades. Go march for peace; go fight for equality; go work for the environment; go feed the hungry; go search for knowledge; or go strive for wealth and position—yet realize that labor for none of these causes will bring as many benefits as labor devoted to making humanity a spacefaring creature. Establishing a self-sufficient and expanding human presence in space is the only way we will elude the nuclear threat, preserve the environment, continue to enjoy democracy, and achieve a standard of living at least equal to that of our parents.

In American history, the young have always been important because it was they who were to conquer new frontiers. So it was in Plymouth in the 1620's, and so it is across the continent in the 1990's. This book is especially directed to those who will make their contributions to our culture in the next fifty years. The free and well-fed youth of America have an opportunity denied to Third World adolescents everywhere, if they could only see what they can achieve. Great movements are afoot. The times call for courage and dedication. Those who have gifts of intellect, passion, or character have even larger obligations. I bid them read these pages with care. Great talents flourish with great objects. Do not lie fallow where you have fallen, but take up the highest cause of your time.

The virtue of ambition has not been respected in recent years. Our novelists and filmmakers are suspicious of those who have visions of commercial success, military glory, political honor, or religious ecstasy. But the era is changing. It is time to raise again the standard of ambition. It is the engine that makes dreams real. It is the force that can call forth the best of what we have within us. Of course, it can be abused, as can any of our better human qualities. Yet we live in difficult times and are faced with hard choices. We need the services of "the granite patience of ambition, which outlasts suffering and shrugs off failure, which retires later than vice and rises earlier than virtue."

We must get off this planet.

A Note to Explorers and Adventurers

In late 1987, a twenty-year-old woman sailed into New York Harbor after completing a two-and-a-half-year solo sail around the world. She may have been young, but she was not unique. Experts estimated that as many as thirty-five hundred people were then circumnavigating the globe. "It's become a milk run," said an author of a book on the subject.

About six months later, an eleven-year-old boy successfully recreated Charles Lindbergh's famous 1927 transatlantic flight. Although his trip was not nonstop and he was accompanied by a retired U.S. Navy commander, the youth was at the controls all the way. At about the same time, a ten-year-old California boy took a diagonal flight across the country, beating the distance record set some months before by another ten-year-old.

In early 1988, thirty-five tourists paid up to thirty-five thousand dollars each to visit the South Pole for several hours. They stopped at a U.S. research installation to have coffee and buy one souvenir apiece. Tourists are not strangers to Antarctica. Over sixty thousand have already visited it. Argentina offers tourists cruises to the area for about twelve thousand dollars per couple and plans a thirty-room hotel at its base camp.

Earth is no longer a place for men who are explorers and adventurers. When boys, young women, and tourists are recreating expeditions that once cost men their lives, we know that true

exploration and adventure have ceased upon this planet. The journeys in the past that excited the world's imagination stretched
human capacity to its limits. Explorers were men who braved the
terrors the unknown holds and lived daily with the possibility of
death far from those they loved. They undertook bold trips for the
sake of fame and the joys of seeing nature's wonders. They strove
to push outward the boundary of human experience. For them,
dangers, privations, and lengthy isolation were commonplace. But
this planet has become a park in which the young, the weak, and
the wealthy imitate the daring of our ancestors. True explorers and
adventurers will be dissatisfied with the meager opportunities left
on Earth. Terrestrial expeditions are too safe and too easy. Explorers and adventurers who want challenges that match their
strength must look off this planet.

The business of explorers is to expand the human domain.
Our species has always been an inhabitant of a world with a def-
inite size and location—the human domain. Surrounding this civilized zone was a frontier; and beyond the frontier were unknown
lands, *terra incognita*. In the beginning, our ancestors inhabited
tribal ground or a walled city or a river valley that was surrounded
by unexplored territory. Later, they occupied regions and continents, until the area of civilization grew to become what popular
thinkers in the mid-twentieth century called a global village. The
human domain now covers the entire planet, and we think of
ourselves as residing on a planet orbiting the Sun and surrounded
by an undifferentiated vastness of space and stars. Our frontier is
now the outer edge of Earth's atmosphere; and the new *terra
incognita* extends beyond. The work of exploration is here.

Four broad regions of exploration and settlement lie before
us. Each lies at a greater distance from Earth than the one before
it; each presents unique problems and opportunities; and each must
be successfully crossed before the next can be attempted. The first
region is cislunar space, from low Earth orbit to the Moon. The
second consists of the inner planets (Mercury, Venus, Mars). In

this region, sunlight is abundant and water scarce. The third is the region of the asteroids and the outer planets (Jupiter, Saturn, Uranus, Neptune, and Pluto). Here, water is abundant and sunlight scarce. The final region is our immediate neighborhood in the galaxy, in which there are five stars within twelve light years that may have planets suitable for life. Given the accomplishments of humanity since 1900, we can establish our presence in each of these regions in the next fifty years.

The purpose of activities in cislunar space will be to establish self-sufficient beachheads for large numbers of people. Mankind will have become truly spacefaring only when either orbiting space cities or lunar colonies have become socially balanced communities exporting something of value to Earth to pay for the supplies they will need. To become a permanent part of human civilization, space colonies must pay their way; otherwise, they will not endure. They must also become true cities, balanced among the sexes, professions, and generations. The Moon itself may one day become an immensely valuable asset. The author Arthur C. Clarke—whose forecasts have been correct before—predicted that in the twenty-first century the Moon will be "more valuable than the wheatfields of Kansas or the oilwells of Oklahoma."

Our objectives in the region of the inner planets will center on Mars. Because of its closeness to the Sun, Mercury may never be more than a scientific outpost or a mining station, similar to the installations now at Earth's poles. Venus is one of the great disappointments of the solar system. Due to the greenhouse effect of its atmosphere, surface temperatures are twice as hot as a standard kitchen oven, and atmospheric pressures are one hundred times greater than they are at Earth's surface. It is unfit for life. Mars, on the other hand, holds great promise. Martian life may never have existed, but Mars itself is the most congenial planet for humans yet discovered. It may one day become mankind's first planetary home beyond Earth.

The asteroids and the outer planets will be the region for the greatest scientific and commercial adventures of the next century. Giant, inhospitable planets, lucrative mining opportunities, and the most spectacular vistas of the solar system are found here. The asteroids are composed of valuable minerals, many of which are becoming scarce on Earth. Some observers believe that the great fortunes of the twenty-first century will be made by entrepreneurs mining the asteroids and shipping the ore back to Earth. Landings and outposts will be on the more habitable moons, of which there are forty-three divided unequally among the five planets. Some of these moons appear very strange, even by the standards of scientists accustomed to the unusual. Europa, one of Jupiter's twelve moons, is nearly featureless and looks like a white billiard ball. One of the strangest objects in the solar system is Iapetus, a moon of Saturn. One side is as black as coal and the other as white as snow. As for commercial possibilities, at least one science-fiction author has written about a new John D. Rockefeller who monopolizes the vast hydrogen reserves known to exist on Titan, another moon of Saturn, to fuel twenty-first-century fusion reactors.

But the great frontier of our times does not lie in the solar system. Consider the meaning of the following list of five stars: Alpha Centauri, Tau Ceti, Epsilon Indi, 61 Cygni, Epsilon Eridani. These oddly spelled, foreign-sounding names denote our nearest neighbors in the galaxy that may have planets suitable for life. All these stars lie within twelve light years of Earth, and their names will one day be familiar to the entire planet, for they will be the initial targets of human efforts to travel beyond this solar system to find worlds suitable for life. Before this generation passes, some few of us will travel to these nearby stars on a lonely and dangerous journey that will rank among the most heroic of mankind's many voyages of discovery. These explorers will never again see Earth's blue skies; they will never again see friends, family, and birthplace; and they will learn to do without many things that cannot possibly be taken on a ship laden with fuel and supplies. And think of the nightmare that may lie at the end of their trip: new planets with surface conditions ferociously hostile to life. Some will wonder how anyone could have the courage to attempt a star trek, yet others will weep bitterly because they are left behind. These starships will be the *Mayflowers* of our time. One day, perhaps within the lifetime of someone of this generation, a starship will head toward one of these suns, containing all of civilization that it can carry, and discover a planet that will become humanity's first outpost beyond the solar system and whose beauty will deserve only one name: New Earth. Its discovery will be an event of galactic significance.

You few who are explorers, lift up your eyes. You have no place here. Planetary frontiers have disappeared in our time, and the solar system surrounding us offers far, far more for exploration than was ever possible on Earth. We who live on this planet at the end of the twentieth century resemble guests arriving too late at a banquet hall after a great feast. We find only the picked-over remains of fabulous dishes and do not realize that our host has planned an even greater meal for us. An age of exploration more marvelous than any that has gone before will begin soon. Wonders beyond belief await you. Don't be left behind.

SOURCE NOTES

PREFACE

7 "one of the pivotal documents": James Miller, *"Democracy Is in the Streets"* (New York: Simon and Schuster, 1987), p. 13. "The Port Huron Statement" is reprinted as an appendix to Miller's book, pp. 329–74. The "an effort in understanding," etc., quotation appears on p. 331.

7 76 million: Landon Jones, *Great Expectations: America and the Baby Boom Generation* (New York: Coward, McCann & Geoghegan, 1980), p. 2.

7 60 percent of electorate: Pat Caddell, "Baby Boomers Come of Political Age," *Wall Street Journal*, 12/30/85. Also, *U.S. News & World Report*, 6/22/87, p. 28.

7 too diverse: For a description of differences, see *Baby Boomers* by Paul C. Light (New York: W. W. Norton, 1988), pp. 75–109.

CHAPTER ONE

19 full-time mother: Jones, op cit., p. 173.

20 two thirds of today's college graduates: Ibid., p. 87.

21 "Never have the young": "The Inheriter," *Time*, 1/6/67, p. 18.

21 Caddell, op. cit.

21 Jones, op. cit., p. 330.

21 American dream: Frank Levy and Richard Michel, "An Economic Bust for the Baby Boom," *Challenge* (May/April 1986), p. 33.

21 "When my family": quoted in Michael X. Delli Carpini, *Stability and Change in American Politics: The Coming of Age of the Generation of the Sixties* (New York: New York

University Press, 1986), pp. 266–67. Appeared first in *New Republic,* 7/9/84, p. 41, attributed to an editor.

22 1980 budget: Jones, op. cit., p.151.

22 fortieth to fiftieth year: Levy and Michel, op. cit., p. 34.

22 A leading Democratic senator: George Will, *Newsweek,* 11/24/86, p. 100. The senator is Daniel Moynihan of New York.

22 Levy and Michel, op. cit., p. 36.

23 worker productivity: Ibid., p. 35. See also Gay Halverson, "US Is Losing Competitive Edge," *Christian Science Monitor,* 5/3/89, p. 9.

23 other causes: Levy and Michel, op. cit., p. 39.

24 Jones, op. cit., p. 174.

24 40 percent of marriages: Ibid., p. 183.

24 couples under thirty: "Here Come the Baby-Boomers," *U.S. News & World Report,* 11/5/84, p. 73.

25 childhood mortality: Jones, op. cit., p. 212.

25 Alexis de Tocqueville, *Democracy in America,* Volume I (New York: Colonial Press, 1899), p. 309.

25 Delli Carpini, op. cit., p. 340.

26 "We must name . . .": quoted in Miller, op. cit., pp. 232–33.

26 "It is the rejection . . .": Delli Carpini, op. cit., p. 326.

27 generation without a party: Ibid., p. 216.

27 "rejects the norms": Ibid.

27 "leaders are disconcertingly hard": Evan Thomas, "Baby-boomers Turn 40," *Time,* 5/19/86, p. 38.

28 "generation may get": Caddell, op. cit.

29 "There is no doubt": quoted in Charles Petit, "A History of Habits," *San Francisco Chronicle, This World,* 2/28/88, p. 7.

29 high school students: Evan Thomas, "Drugs, the Enemy Within," *Time,* 9/15/86, pp. 59–68.

29 cocaine use: "Big Drop in Cocaine Use Among High School Seniors," *Los Angeles Times* News Service, printed in *San Francisco Chronicle* 1/14/88. Drop was from 12.7 percent in 1986 to 10.3 percent in 1987.

30 AIDS: Jeffrey Harris, "The AIDS Epidemic: Looking Into the 1990's," *Technology Review,* 6/87, pp. 58–64.

32 extinction of mankind: Jonathan Schell, *The Fate of the Earth,* (New York: Alfred A. Knopf, 1982), p. 65. The title of Chapter One is "A Republic of Insects and Grasses."

34 megacities: Robert McNamara, "Time Bomb or Myth: The Population Problem," *Foreign Affairs* (Summer 1984), p. 1117.

35 "a condition": Ibid., pp. 1118–19.

36 cropland growth: Lester Brown, et al., *State of the World 1985: A Worldwatch Institute Report on Progress Toward a Sustainable Society* (London and New York: W. W. Norton), p. 24.

37 fertilization and irrigation: Ibid., p. 22.

37 fisheries: Ibid., p. 76.

37 1984 analysis: William Hartman, Ron Miller, and Pamela Lee, *Out of the Cradle: Exploring the Frontiers Beyond Earth* (New York: Workman Publishing, 1984), p. 34. The 1984 analysis is a study by Oak Ridge Laboratory technologists H. E. Goeller and A. Zucker published in *Science.*

37 Bangladesh: Brown, op. cit., p. 17.

39 worsening problems: Peter G. Peterson, "The Morning After," *Atlantic Monthly,* 10/87, pp. 43–69.

40 2030: Lee Smith, "The War Between the Generations," *Fortune,* 7/10/87.

40 1986 government spending: Peterson, op. cit., p. 60.

41 "We face": Ibid., p. 64.

41 "The assumption": Smith, op. cit.

CHAPTER TWO

43 Gaston's: Frank Viviano, "California," *San Francisco Sunday Examiner and Chronicle, Image,* p. 6.

44 "of the living": Arnold Toynbee, *A Study of History,* Vol-

ume 9 (New York: Oxford University Press, 1954), p. 411.

44 "most thoroughgoing": William McNeill, *The Rise of the West* (Chicago: University of Chicago Press, 1963), p. 668. In this passage, McNeill is referring specifically to the New England and Middle Atlantic colonies of North America in the eighteenth century.

47 *Homo* evolution: Jay Greene, the Institute of Human Origins, Berkeley, California, in private correspondence to Deborah Nikkel, January 10, 1989. Mr. Greene confirmed these facts with the anthropologists Yoel Rak and Bill Kimbel.

50 "From the perspective": McNeill, op. cit., p. 567.

50 "apparently in the" and "the distorting": Toynbee, op. cit., p. 412.

50 egocentric illusion: Ibid., p. 410.

53 "the golden age": Carl Sagan, *Broca's Brain* (New York: Random House, 1979), pp. 213–14.

53 "the present moment" and "not many generations": Carl Sagan, *Cosmic Connection* (New York: Doubleday, 1973), p. 69.

56 Jedediah Smith: Allan Nevins and Henry Steele Commager, *A Short History of the United States* (New York: Alfred A. Knopf, 6th edition, 1984), p. 213.

57 Oregon Trail: Ibid., pp. 214–15.

57 space society: This group is called the L-5 Society. In 1988, it merged with the National Space Institute to become the National Space Society.

61 entrepreneur subculture: For examples, see any issue of *Inc.*, *Venture,* or *Success.*

61 incubators: Kerry Elizabeth Knobelsdorff, " 'Incubators' help nurture good ideas into profitable ventures," *Christian Science Monitor,* 11/10/87, p. 11.

61 franchises: Janice Castro, "Franchising Fever," *Time,* 8/31/87, pp. 36–38. Also, "Franchising Grows from Burgers to Boutiques," by Kerry Elizabeth Knobelsdorff, *Christian Science Monitor,* 2/29/88, pp. 16–17.

61   Silicon Valley and Route 128: For example, on September 26, 1987, the San Jose State School of Business sponsored a daylong series of classes entitled "Insight on Entreprenurial Success." For seventy-five dollars, the student could "cover the full range of factors directly effecting the ultimate success" of his business venture. Such events are common.

62   "a wise plan": Nevins and Commager, op. cit., p. 118.

65   constitutions: Walter Berns, *Taking the Constitution Seriously* (New York: Simon and Schuster, 1987), p. 12.

65   "From its earliest": Nevins and Commager, op. cit., v.

66   "We Americans" and "I, for one,": quoted in Arthur Schlesinger, Jr., *Cycles of American History* (Boston: Houghton Mifflin, 1986), pp. 15–16.

66   "We are destined": Richard Reeves, "The Meanest Inaugural Address," *San Francisco Chronicle*, 1/24/85, p. 57.

67   imperial variant of Manifest Destiny: Daniel J. Boorstin, *The Americans, the National Experience* (New York: Random House, 1966), p. 271; and John Carl Parish, "The Emergence of the Idea of Manifest Destiny," UCLA Faculty Research Lectures, 5/6/31.

67   "A free, confederated": Frederick Merk, *Manifest Destiny and Mission in American History* (New York: Alfred A. Knopf, 1963), p. 29.

68   "I took Panama": quoted in Nevins and Commager, op. cit., p. 420.

69   "Our manifest destiny": quoted in Merk, op. cit., p. 32.

CHAPTER THREE

72   "It is not easy": Schlesinger, op. cit., p. 6–7.

73   oldest democracy: Nevins and Commager, op. cit., v. "[America] is today the oldest republic and oldest democracy and lives under the oldest written constitution in the world."

75   "In 1842, a man named": Boorstin, op. cit., frontispiece.
76   Fn "Even Turner's": Richard Hofstadter, *The Progressive Historians* (Chicago: University of Chicago Press, 1968), p. 119.
77   "the repeated rebirth": Ray Allen Billington, *Westward Expansion,* 3d edition (New York: Macmillan Company, 1950), p. 1.
78   new social order: Hofstadter, op. cit., p. 155.
78   agent of change: Avery Craven, "Frederick Jackson Turner," 1937. Reprinted in *The Turner Thesis,* edited by George Rogers Taylor (D. C. Heath & Co., 1956).
78–79   frontier and democracy: Stanley Elkins and Eric McKitrick, "A Meaning For Turner's Frontier: Democracy in the Old Northwest," 1954. Reprinted in *Turner And the Sociology of the Frontier,* edited by Richard Hofstadter and Seymour Martin Lipset (New York: Basic Books, 1968), pp. 120–44.
81   frontier and labor: George G. S. Murphy and Arnold Zellner, "Sequential Growth, the Labor Safety-Valve Doctrine and the Development of American Unionism," 1959. Reprinted in Hofstadter and Lipset, eds., op. cit., pp. 201–224. Also, see Hofstadter, *The Progressive Historians,* p. 160—"We must think of the West as a magnificent area for the rapid expansion of middle-class society."
82   reformers and immigrants: Oscar Handlin, *The Uprooted* (Boston: Little, Brown, 1951), pp. 218–19.
82   natural leaders: Crane Brinton, *Anatomy of a Revolution* (New York: W. W. Norton, 1938), p. 75—"Able men do seem to get born in the humblest ranks, and an accumulation of able and discontented men would provide splendid natural leaders for groups restive and ready for revolt."
83   Emerson's typical Yankee: Nevins and Commager, op. cit., p. 205.

83   "For three centuries": Frederick Jackson Turner, *The Frontier in American History* (New York: Henry Holt & Company, 1920), p. 293.

84   multimillionaires: Jacqueline Thompson, *Future Rich* (New York: William Morrow, 1985), pp. 13–14.

86   rich becoming richer: Barbara Ehrenreich, "Two Americas," *This World, San Francisco Sunday Examiner and Chronicle*, 11/23/86, p. 9.

86   divided America: James Fallows, *More Like Us: Making America Great Again* (Boston: Houghton Mifflin, 1989), p. 175. Although Fallows believes that "there is still no conclusive proof that America's middle class has shrunk," he does agree that "in cultural terms, America is much less middle class than it was a generation ago." See also: James Lardner, "Rich, Richer; Poor, Poorer," *New York Times* 4/18/89, p. A19—"We have begun to entertain the idea that severe and lasting material differences among people—a class system, in short—is the natural condition of humanity"; and Laurent Belsie, "Poor Are Getting Poorer in the US," *Christian Science Monitor*, 5/14/89, p. 8.

86   Daniel Bell: Rushworth Kidder, "State of the World in 2013—According to Daniel Bell," *Christian Science Monitor*, 9/28/87, p. 19.

87   Los Angeles freeways: Kevin Roderick, "Life in the Fast Lane," *This World, San Francisco Sunday Examiner and Chronicle*, 5/10/87, pp. 9–10. Article appeared originally in *Los Angeles Times*.

88   "a boundless continent" etc.: de Tocqueville, op. cit., pp. 294–95.

88   "This then is": Turner, op. cit., p. 358.

89   "has spread": quoted in Schlesinger, op. cit., p. 119.

90   "can doubt": Ibid., p. 141.

92   war deaths: James Dunnigan and Austin Bay, *A Quick and*

Dirty Guide to War (New York: William Morrow, 1985), pp. 320–24; and *The World Almanac 1988*, "Casualties in Principal Wars of the U.S.," p. 338.

92 "persevere in design," etc.: de Tocqueville, op. cit., p. 239.

92 "open covenants": quoted in Nevins and Commager, op. cit., p. 454.

94 Atlantic Council: Douglas MacArthur II, "Presidents Need to Be Global in Outlook," *Christian Science Monitor*, 2/10/87, p. 18.

95 challenge of China: "China is so large and its economic growth rate so strong that by 2010, RAND [Corporation in a new study] predicts, it could even pass Japan and become the second largest economy in the world, after the United States." Peter Grier, "World Power Center Shifts Toward Pacific Rim, Study Says," *Christian Science Monitor*, 5/17/89.

96 China's economic growth: Dwight Perkins, *China: Asia's Next Economic Giant?* (Seattle: University of Washington Press, 1986), p. 85.

97 Chinese agriculture: "Setting a Full Table," *Time*, 10/12/87, p. 55.

97 "Here they have": Kenneth Scott Latourette, *The Chinese: Their History and Culture*, 4th edition (New York: Macmillan, 1964), p. 450.

97 "outstanding merits": Jacques Gernet, *A History of Chinese Civilization* (Cambridge and New York: Cambridge University Press, 1982), p. 27.

98 "The big news": Norman Cousins, *Christian Science Monitor*, 11/12/86, pp. 28–29.

101 "He would be": Turner, op. cit., p. 37.

101 "The opening": Gerard K. O'Neill, *The High Frontier: Human Colonies in Space* (New York: William Morrow, 1977), p. 232.

CHAPTER FOUR

104   "extraterrestrial imperative": O'Neill, op. cit., p. 258.

106   "employed as a discoverer": Ben R. Finney and Eric M.
Jones, "The Exploring Animal," *Intersteller Migration and
the Human Experience* (Berkeley, Los Angeles, London:
University of California Press, 1985), p. 22.

107   "it is entirely": Robert Rood and James Trefil, *Are We
Alone? The Possibility of Extraterrestrial Life in the Universe* (New York: Charles Scribner's Sons, 1981), p. 254.
Rood and Trefil are concerned with the existence of "advanced races," which they define as intelligent life capable
of developing advanced communications technology. They
use the so-called Greenbank Equation to calculate that the
chances are only about 1 in 300 (0.3 percent) that any civilization is communicating in any given year in the galaxy.
Using their estimates most favorable to the existence of extraterrestrial life, the chances in any given year of intelligent
life existing off Earth is about four times that value or 1 in
100 (4 x .003 = 1.2 percent).

107   Ibid., p. 255.

107–109   intelligent life off Earth: In addition to Rood and Trefil,
consult "Are We Alone?" by Gregg Easterbrook in the
Atlantic Monthly, 8/88, pp. 25–38; and *Spaceships of the
Mind*, by Nigel Calder (Westford, Massachusetts: Penguin
Books, 1979), p. 137.

111   "More than any other man": quoted in Nevins and Commager, op. cit., pp. 142–43.

112   "to the hundredth and thousandth generation": Ibid.,
p. 291.

113   "passionately in the ability": Freeman Dyson, *Disturbing
the Universe* (New York: Harper Row, 1979), p. 118.

114   O'Neill's influence: For example, see Michael Michaud,

Reaching for the High Frontier: the American Pro-Space Movement 1972–84 (New York, London, and Westport, Connecticut: Praeger, 1986), p. 60; Louis Halle, "A Hopeful Future for Mankind," *Foreign Affairs,* pp. 1132–36; *Space Colonies,* edited by Stewart Brand (San Francisco: Penguin Books, 1977); and Daniel Deudney, *The High Frontier in Perspective* (Washington, D.C.: Worldwatch Institute, 1982).

115   west of Missouri: Nevins and Commager, op. cit., pp. 338–39.

115   8 million immigrants: Ibid., pp. 351–55.

116   "O'Neill and I": Dyson, op. cit., p. 126.

117   "The general impression": op. cit., p. 70.

118   "The engineering aspects": quoted in Michaud, op. cit., p. 70.

119   "The typical letter": O'Neill, op. cit., p. 252.

122   "conceptual engineering studies": The National Commission on Space, *Pioneering the Space Frontier* (New York: Bantam Books, 1986), p. 72.

124   "When we consider": O'Neill, op. cit., p. 224.

125   "it would require": Ibid.

125   airline passengers: U.S. Department of Commerce, *Statistical Abstract 1987,* p. 603; also, *Historical Statistics of the United States, 1789–1970* (Washington, D.C.: U.S. Government Printing Office, 1975), Part 2, p. 769.

128   "For today": Schell, op. cit., p. 224.

CHAPTER FIVE

132   "as a truly extraordinary": William H McNeill, "The Dream of World Culture," on p. 172 of *Twentieth Century,* by Joel G. Colton and the editors of Time-Life Books, 1968.

134   "man will stand": quoted in Dyson, op. cit., p. 226.

135 voting: U.S. Department of Commerce *Statistical Abstract of the United States, 1988* (Washington, D.C.: U.S. Government Printing Office, 1975), p. 250. Figures for the 1988 presidential election were taken from *The New York Times*, 3/12/89, p. 22. The 1988 turnout of 50 percent was the lowest since 1924.

136 "Today, Republicans hold": George Gallup, "Democrats Still Have More Party Members," *San Francisco Chronicle*, 10/19/87, p. A9.

136 "no longer embodies": Robert Reich, "Towards a New Public Philosophy," *Atlantic Monthly*, 5/85, p. 71.

136 "less the embodiment": Ibid., p. 72.

137 a new public philosophy: Ibid., pp. 78–79.

138 "an infinity" and "of a special destiny": Nevins and Commager, op. cit., vii.

139 "attractive settlements": National Commission on Space, p. 72.

140 NASA report: Michael Simon, *Keeping The Dream Alive: Putting NASA and America Back in Space* (San Diego: Earth Space Operations, 1987), p. 179.

141 Wright Brothers: "Heads in Air, Feet on Ground," *Time*, 6/1/87, p. 68.

142 "the most exciting": Dyson. op. cit., p. 114.

142 "The history": Ibid., p. 115.

143 "A small group": Ibid., p. 110.

143 "I believe": Ibid., p. 117.

143 "To send troops": quoted in J. S. Holliday, *The World Rushed In: The California Gold Rush Experience* (New York: Simon and Schuster, 1981), pp. 35–36.

144 San Francisco population: Charles Lockwood, *Suddenly San Francisco* (San Francisco: The Hearst Corporation, 1978), p. 73. In 1853, San Francisco has a population of fifty thousand—eight thousand of whom were women.

144 National Space Society: See also *Spacefaring Gazette, A*

Journal for Space Development, published by the Golden Gate Chapter of the National Space Society (4009 Everett Avenue, Oakland, CA 94602). The regional council is the California Space Development Council, which adopted "this statement of vision" at their first conference in 1985.

145 "I believe": Peter Glaser, "Business in Space," *Atlantic Monthly,* 5/85, p. 45.

146 "In the long run": National Commission on Space, op. cit., p. 88.

146 natural resources in solar system: John S. Lewis and Ruth A. Lewis, *Space Resources: Breaking the Bonds of Earth* (New York: Columbia University Press, 1987), p. 4. "Every raw material that we could ever want is up there, often in quantities and concentrations beyond anything available on Earth."

148 "Prospecters and miners": Ben Bova, *The High Road* (New York: Pocket Books, 1983), p. 187.

149 "historians of the future": William D. Nixon and Richard E. MacCormack, "Landsat: A Tool for Your Classroom," *Social Education,* vol. 41, no. 7, Nov./Dec. 1977. Cited in "A Current Perspective on Space Commercialization" (Washington, D.C.: The Aerospace Research Center, 1985), pp. 17–18.

150 payload costs: Lewis and Lewis, op. cit., p. 160.

150 change structure of space program: Phillip K. Salin, "Toward a New U.S. Space Transportation Policy," presentation to National Commission on Space Public Hearings, San Francisco, 11/14/85.

151 "Almost all the nations": de Tocqueville, op. cit., p. 257.

151 "to constitute a people": Ibid., p. 239.

152 Soviet time in space: "The Soviets in Space," *U.S. News & World Report,* 5/16/88, p. 53.

153 *Energeia*: Jerry Grey, *Beachheads in Space: A Blueprint for the Future* (New York: Macmillan, 1983), pp. 34–46. Also,

Michael Lemonick, "Surging Ahead," *Time,* 10/5/87, pp. 64–69.

153 "The Soviet Union": Bova, op. cit., p. 32.

154 China: Finney and Jones, op. cit., pp. 201–202.

156 changes in space policy: Jerry Grey, *Government v. Free Enterprise in Space* (Dallas: National Center for Policy Analysis, 1987, Executive Summary.)

157 "the institutional structure": *Reducing Launch Costs: New Technologies and Practices* (Washington, D.C.: Office of Technology Assessment, Congress of the United States, 1988), pp. 6, 8.

159 heroes: J. S. Holliday, "Have Celebrities Replaced Heroes?" Speech to the Commonwealth Club of California, 8/12/88.

161 "Perhaps no form": quoted in Schlesinger, op. cit., p. 429. To which Schlesinger adds, "Leadership—the capacity to inspire and mobilize masses of people—is a public transaction with history."

161 "You really have," quoted in "Playing Politics by the Numbers," *U.S. News & World Report,* 6/22/87, p. 29.

161 Max Weber, *Essays in Sociology* (New York: Oxford University Press, 1946), p. 128.

CHAPTER SIX

164 spending and labor force: "Here Come the Baby-Boomers," *U.S. News & World Report,* 11/5/84, p. 68; and Lynn Rosellini, "When a Generation Turns 40," Ibid., p. 60.

164 AARP: Nancy R. Gibbs, "Grays on the Go," *Time,* 2/22/88, p. 69.

165 population older than sixty-five: John Tierney, "Old Money, New Power," *New York Times Magazine,* 10/23/88. An excellent essay on AARP's influence today.

165 living to be eighty-five: Jones, op. cit., p. 314.

165 "within the next 20 years": "Longer Lives for 'Senior Boomers,' " *San Francisco Chronicle,* 10/31/87, p. A3.

165 people living past eighty-five: Cheryl Russell, *100 Predictions for the Baby Boom: The Next 50 Years* (New York: Plenum Press, 1987).

165 share of population: Gibbs, op. cit., pp. 66–79.

167 "We ended a war," quoted in Miller, op. cit., p. 16.

167 "the perimeter": Ibid., p. 367.

168 "we were walking": Ibid., pp. 57, 60.

168 excitement of 1960's: See also Todd Gitlin, *The Sixties: Years of Hope, Days of Rage* (New York: Bantam Books, 1987).

168 "A Mass Movement": Miller, op. cit., p. 328. These are the final words of Miller's book.

173 "Americans . . . probably.": Nevins and Commager, op. cit., p. 294.

173 Technology will not: Rushworth M. Kidder, "Freeman Dyson," *Christian Science Monitor,* 10/14/86, pp. 28–29.

174 thirty-year cycles: Schlesinger, op. cit., pp. 23–48.

CHAPTER SEVEN

178 humanity's great ages: McNeill, op. cit., p. 807—"Men some centuries from now will surely look back upon our time as a golden age of unparalleled technical, intellectual, institutional and perhaps even of artistic creativity. We . . . should count ourselves fortunate to live in one of the great ages of the world."

179 "This is the American moment": Allan Bloom, *The Closing of the American Mind* (New York: Simon & Schuster, 1987), p. 382. This is the first sentence of the final paragraph of Bloom's book.

180 "[Dr. Gerard] O'Neill": Dyson, op. cit., p. 126. In "A Space Traveler's Manifesto" (July 1958), Dyson wrote the following memorable passage: "It is in the long run essential to the growth of any new and high civilization that small groups of people can escape from their neighbors and from their governments, to go and live as they please in the wilderness. A truly isolated, small and creative society will never again be possible on this planet." Excerpted in *Disturbing the Universe*.

182 "the granite patience": Ted Morgan, *Churchill, Young Man in a Hurry* (New York: Simon & Schuster, 1982), p. 564. These phrases are from the final paragraph in Morgan's book.

APPENDIX

183 "It's become": Ron Scherer, "Globe-circling sailors pursue the ultimate getaway," *Christian Science Monitor*, 11/5/87, p. 6.

184 For a comparison between the voyages of discovery in the fifteenth and sixteenth centuries and interplanetary journeys in our time, see Carl Sagan's *Cosmos* (New York: Random House, 1980), p. 139.

184–185 the planets: For a description of the planets and their moons, see *Out of the Cradle* (cited in Chapter One) and also *The Far Planets*, Time-Life Books, "Voyage Through the Universe."

185 "more valuable": quoted in Adrian Berry, *The Next Ten Thousand Years* (New York: New American Library, 1974), p. 46. Clarke's statement originally appeared in a magazine article that was reprinted in his *The Promise of Space* (1968).

186 nearby stars: Ken Croswell, "Looking for a Few Good Stars," *Space World*, 7/87, pp. 12–15.

INDEX

AARP (American Association of
   Retired Persons), 164–165
acid rain, 38, 176
Aegean civilization, 51
Aeschylus, 99
Afghanistan, 81
Africa, 44, 48, 56, 89, 102, 106
   AIDS in, 30
   colonization of, 43, 49, 59
   food supply of, 35
AIAA (American Institute of Aero-
   nautics and Astronautics), 120
AIDS (acquired immune deficiency
   syndrome), 26, 30
alcohol, teenagers' abuse of, 30
Alexander the Great, 48–49
Alliance for Aging Research, 165
Alpha Centauri, 186
aluminum, dwindling supply of, 37
American Association of Retired
   Persons (AARP), 164–165
American Commonwealth, The
   (Bryce), 161
American Enterprise Institute, 65
American Institute of Aeronautics
   and Astronautics (AIAA), 120
Amundsen, Roald, 52, 56
Antarctica, 52–53, 56, 183
   exploration of, 52–53, 56
   tourism in, 183
Apache Indians, 115
Apollo program, 155
   cost of, 121, 156
   in history, 54–55
   knowledge gained through, 118
   success of, 53, 59, 152, 173, 174
Appalachian Mountains, 77
Apple Computer, 146
Arabic language, 51
Argentina, 53, 183
Aristotle, 99

Arizona, 68
Arkansas, 77
Ashley, William, 56
Asia, 49, 59
asteroids, as mineral resource, 124,
   145, 146, 148, 185–186
Astor, John Jacob, 83
Athens, U.S. resemblance to, 99–
   100
Atlantic Council, 94
Australia, 51, 53, 72, 89
Aztecs, 48

Baby and Child Care (Spock), 19
Baikonur Cosmodrome, 152–153
Baltimore Sun, 128
Bangladesh, 34, 37
Barbie dolls, 164
Becker, Carl, 111
Bell, Daniel, 28, 86
Bhopal (India), 38, 176
Billington, Ray Allen, 77
Bill of Rights, U.S., 18
binary stars, 108
Blackfoot Indians, 115
Blacks:
   immigrants and, 87
   Republican party and, 135
Bloom, Allan, 179
Bloomingdale's, 86
Boston, Mass., 61
Bradford, William, 114
Brandeis, Louis, 159
Brazil, 49, 53, 98
British Commonwealth of Nations,
   89
Bryan, William Jennings, 82
Bryce, James, 161
Buckley, William F., Jr., 28
Buddhism, 49, 51
buffalo, 115